# William Whewell

# Theory of Scientific Method

ISBN 0-87220-083-3 cloth
ISBN 0-87220-082-5 paperback

Originally published 1968 by *University of Pittsburgh Press*
LOC Catalog Card Number 68-23166

# William Whewell
## Theory of Scientific Method

---

*Edited*
*with an Introduction*
*by*

**Robert E. Butts**

*Hackett Publishing Company*
*Indianapolis/Cambridge*

Printed in the United States of America

A reprinting, with revisions, of the 1968 edition

Cover design by Listenberger Design and Associates

For further information, please address

Hackett Publishing Company, Inc.
P.O. Box 44937
Indianapolis, Indiana 46204

**Library of Congress Cataloging-in-Publication Data**

Whewell, William, 1794–1866
William Whewell's theory of scientific method
edited with an introduction by Robert E. Butts
p. cm.
Reprint. Originally published: Pittsburgh:
University of Pittsburgh, 1968.
Bibliography: p.
Includes index.
ISBN 0-87220-083-3 ISBN 0-87220-082-5 (pbk.)
1. Science—Philosophy. 2. Science—Methodology.
I. Butts, Robert E. II. Title. III. Title:
Theory of scientific method.
Q175.W567 1989 88-37200
501—dc19 CIP

The paper used in this publication meets the minimum requirements of American National Standard for Information Sciences—Permanence of Paper for Printed Library Materials, ANSI Z39.48-1984. ∞

*For*
*Eugènie-Valentine*

# *Preface*

When the first version of this book appeared in 1968, the philosophy of science of William Whewell was known by few. In preparing the book, it had been my intention to bring Whewell's neglected theories to the attention of philosophers of science and their students, in part to provide an alternative interpretation of science. Whewell attempted to correct what he thought were mistakes in John Stuart Mill's theory of induction. Mill carried on the debate in later editions of his *System of Logic*. Indeed, it was hoped that this anthology could be used in seminars along with Mill's *Logic*, as I have done in my own teaching. Now that the philosophical scene has changed, perhaps that early hope can be better fulfilled by means of this reissue of the work.

The only earlier major figures to recognize the importance of the debate were Charles Sanders Peirce and Ernst Mach. Now we see an important new form of the Whewell/Mill exchange in the debate between the antirealists (empiricists) and the realists (rationalists) in philosophy of science, with Bas van Fraassen leading the way for the empiricists, and Michael Friedman and others championing a more Whewellian reading of the nature of science. On the empiricist side, what is claimed is that we can settle for theories that are empirically adequate, with Whewellian considerations of simplicity and consilience playing only pragmatic roles. On the realist side, equally strong claims are made on behalf of theoretical unification and consilience of inductions as epistemically definitive of the best theories in science. What is to be noted is that amongst philosophers of science today 'Whewell' and 'consilience' have become household words. More than that: some of the most important work on inductive methods and theory of confirmation has now begun to take the work of Whewell seriously.

The selections in the volume are meant to focus attention on Whewell's theory of scientific method. I will argue in the Introduction that Whewell's philosophy of science stresses an important concept of necessary truth (selections in Chapters I and II) and an equally important concept of induction (selections in Chapters III, IV, and V). The selections are arranged in such a manner that Whewell's continuing arguments for his central theses are easily apparent.

I have abridged some of the selections, but hope not to have damaged the continuity of the arguments in doing so. Whewell's spelling, use of capital letters, and punctuation, though not always consistent, have been retained throughout. Whewell's footnotes, in the main, are abbreviated; some are not very helpful, and some are simply inaccurate. I have therefore filled out the abbreviated notes, eliminated some others, and tried to correct mistakes. I have, however, retained all of Whewell's cross-references to his own works. Only Whewell's explanatory footnotes have been retained in their original form. Republication of the work has given me the opportunity to correct a mistake in and to rewrite some parts of the original Introduction and to update the Bibliography.

The debts I owe to others have grown larger in the twenty years since this work was first published. Henry Hiz first suggested that I study Whewell. Dr. Robert Robson, Fellow of Trinity College, Cambridge, sponsored my use of the papers of Whewell in 1962–63. Since then, those who have taught me about Whewell cannot properly be thanked. I must mention Gerd Buchdahl, Mary Hesse, Larry Laudan, William Harper, James Brown, Kathleen Okruhlik, Joseph Pitt, Michael Ruse, and David Wilson.

I am grateful to Jay Hullett of the Hackett Publishing Company for his interest in republication of this anthology. The opportunity to work with the people at Hackett has been refreshing and rewarding.

Both books and scholars can be reborn. Eugènie-Valentine has made the difference. This book is for her.

ROBERT E. BUTTS
London, Ontario, Canada
Halloween 1988

# *Contents*

**William Whewell**

# *Theory of Scientific Method*

# *Introduction*

## BIOGRAPHICAL NOTE

William Whewell was born in Lancaster, England, in 1794, and died in Cambridge in 1866. His life was one of intensive and extensive work in science, scholarship, and academic administration. In a sense, his life began when he entered Trinity College, Cambridge, in 1812. From that point on, his energies and interests were almost totally bound up with the life of his College and of the University. He was quickly identified as the best scholar in his year at Trinity. Election as a Fellow of the College came in 1817; appointment as Master followed in 1841. In the University, he held professorships first in mineralogy, then in moral philosophy. He was Vice-Chancellor of the University in 1842 and again in 1855. During the last twenty-five years of his life he married twice.

Whewell had little time for anything but scientific and scholarly work. He did important empirical work on the tides, produced expert books and papers in physics, geology, economics, astronomy, and architecture; he also found time to compose some poetry, sermons, and theological pieces. In 1837 he published one of his two most important books, the *History of the Inductive Sciences.* This book was the first full-scale history of empirical science, and still claims a place as a useful study. In 1840 his second major work appeared, *The Philosophy of the Inductive Sciences, founded upon their History.* The work provoked great controversy, and was reprinted twice in the twenty years following its first appearance. In the third edition (1858–1860), it appeared as three separate but related works: *History of Scientific Ideas, Novum Organon Renovatum,* and *The Philosophy of Discovery.*[1]

## THE ELEMENTS OF WHEWELL'S PHILOSOPHY OF SCIENCE

Whewell's general philosophy of science is in many features strikingly modern. He was concerned, but not primarily, with what is now referred to as the logic of science, an explication of the detailed formal connections among parts in finished scientific systems. He also tried to formulate general rules of methodology, partly in the interest of producing rules that could actually be followed by subsequent scientists, partly because such methodological generalizations account for some of the logic of science. He also attempted to work out those epistemological issues so important to contemporary philosophers of science, issues generally concerned with the strength and credibility of scientific conclusions. Indeed, all three of these interests resulted in a slogan that most recent philosophers of science would accept: "Man is the Interpreter of Nature, and Science is the right Interpretation." Of course the task of the philosopher of science, as Whewell viewed him, was to make this slogan stick as a philosophical truth.

Nevertheless, for all of its modernity, Whewell's philosophy of science also exhibits anachronistic elements. He thought of scientific systems as resulting in necessary laws, whose necessity could not be accounted for merely by appealing to the logical necessity characteristic of the logico-mathematical form of those sciences. Coupled with this idea was his confident belief that inductive inferences are demonstrative, which meant for him that the conclusions of some inductive inferences are necessary. Finally, Whewell cast his philosophy of science in the larger framework of a metaphysical theory, part of which was based on Christian theological principles, part on the quasi-Hegelian identification of concepts that are usually viewed as separate (e.g., his argument that theories and facts, ideas and sensations are *not* really separate), and part on dubious psychologistic and genetic prejudices that stemmed largely from his understanding of progress in the history of science.

The central, and clearly the most intriguing, thesis of Whewell's philosophy of science is that science develops by becoming a more and more comprehensive system of laws that are both universal and necessary, and that are, nevertheless, in some sense the result of induction. The thesis rings strangely in ears accustomed to recent controversy over induction and theory of probability; it also rang strangely in nineteenth-century ears familiar with the damaging criticisms of induction evinced by Hume. But the thesis is well worth studying as the core of one classical English philosophy of science. The selections that follow in this volume were chosen to illustrate both the thesis itself and Whewell's method of defending it. They are arranged in such a way that we may see Whewell as arguing for a connected set of propositions in both methodology and epistemology, the arguments eventually resulting in the richly-formulated theory of scientific method contained in *Novum Organon Renovatum* and in the reply to Mill. The selections, of course, speak for themselves, and the job of the editor is done if a rational organization has been achieved. However, since the concepts of necessary truth and induction are so vital to Whewell's defense of his thesis, I shall endeavor to supply some broader context for the understanding of these concepts in what remains of this introduction.

## WHEWELL'S THEORY OF NECESSARY TRUTH

Whewell's theory of necessary truth involves three major propositions.[2] (1) Whatever sciences there are (as sciences) contain what Whewell calls "Fundamental Ideas" that are the sources of necessary truths (sometimes called "Axioms"), certain ideal relations on which the very existence of those sciences depend. (2) These necessary truths have reference to empirical realities that, in the course of experience, we know *before* we apprehend as necessary truths the propositions that express these realities. (3) We see intuitively that propositions are necessary when we cannot clearly and distinctly conceive their logical contraries. Whewell's arguments for the three propositions constitute what might be called his answer to the Kantian question: "How are necessary and universal truths possible?"

For Whewell, the Fundamental Ideas "are not Objects of Thought, but rather Laws of Thought. Ideas are not synonymous with Notions; they are Principles which give to our Notions whatever they contain of truth."[3] He also states that "by the word Idea (or Fundamental Idea) used in a peculiar sense, I mean certain wide and general fields of intelligible relation, such as Space, Number, Cause, Likeness."[4] Fundamental Ideas are what the activity of mind contributes to knowing. Whewell likens some of them, notably space, time, and number, to Kant's forms of intuition. Others, for instance the ideas of cause and likeness, play for Whewell something akin to the role of Kant's categories, though he does not use Kant's term to designate them. Futhermore, in its treatment of some of the Fundamental Ideas, especially space and time, Whewell's account of their epistemological status deviates very little from the Kantian theory.[5] Thus the Ideas of space and time inform our sensational experience (without being derivable from it), making meaningful perception possible. Whewell also speaks of Ideas as subjective forms for interpreting experience in such a way that knowledge-yielding statements about it become possible. Finally, the philosophy of each science consists in the development of the Fundamental Ideas that articulate and organize the propositions of that science, and give it whatever in the nature of truth status it might possess.

So regarded, Whewell's Fundamental Ideas are simply Kant's forms of intuition and categories under a new name. What is significant for our purposes is not this evident similarity of his doctrine to Kant's (which Whewell readily admits), but rather the novel features of Whewell's position, which his subsequent discussion brings forth. For Whewell's central (and largely novel) contention is this:

> The Progress of Science consists in a perpetual reduction of Facts to Ideas . . . Necessary Truths belong to the Subjective, Observed Facts, to the Objective side of our knowledge. Now in the progress of that exact speculative knowledge which we call Science, Facts which were at a previous period merely Observed Facts, come to be known as Necessary Truths; and the attempts at new advances in science generally introduce the representation of known truths of fact, as included in higher and wider truths, and therefore, so far, necessary. . . . Such steps in science are made, whenever empirical facts are discerned to be necessary laws; or, if I may be allowed to use a briefer expression, whenever *facts are idealized.*[6]

This view that empirically observed truths can become necessary ones, or as Whewell says elsewhere, that a posteriori truths become a priori[7] appears to be quite incompatible with the Kantiansim of his general conception of the Fundamental Ideas. On the one hand Whewell holds that there is a distinction between necessary and empirical truths, and thus that necessary truths cannot depend for their evidence upon appeal to experience; on the other hand, he believes that necessary truths emerge as necessary in the course of the development of this or that empirical science. But if the latter is the case, it is difficult to see how necessary truths can be rigorously distinguished from empirical truths, and how they can be, as in the case of space and time, conditions residing in the constitution of the human mind, to which all present and all future experiences must conform.

To understand this initially astonishing view, one must comprehend in detail both the nature of Whewell's Fundamental Ideas and the character of necessity that they bestow on some propositions of fact. It is clear that, for Whewell, there are at least as many Ideas as there are sciences, and that each set of Ideas relative to a given science makes possible the expression of the laws of that science. But since sciences develop in concrete historical situations and over long periods of time, it follows that we do not now know every Idea that there is to be known. Thus Whewell's theory does not imply that the mind is pre-stocked with such Ideas and is therefore ready at once to develop particular sciences. For Whewell, quite the contrary seems to be the case.

> It is not the *first*, but the most complete and developed condition of our conceptions which enables us to see what are axiomatic truths in each province of human speculation. Our fundamental ideas are necessary conditions of knowledge, universal forms of intuition; inherent types of mental development; they may even be termed, if any one chooses, results of connate intellectual tendencies; but we cannot term them *innate* ideas, without calling up a large array of false opinions. . . . Fundamental Ideas, as we view them, are not only not innate, in any usual or useful sense, but they are not necessarily *ultimate* elements of our knowledge. They are the results of our analysis so far as we have yet prosecuted it; but they may themselves subsequently be analysed.[8]

The position seems, therefore, to be something like this: Man is subjected to a wide variety of sensations, over many of which he exerts no form of control, and which constitute the "matter" of knowledge. But these sensations, in order to yield knowledge in the form of general propositions, must be formed, and each of the forms that we actively bestow upon experience becomes a candidate for scientific necessary truth. It is clear, however, that not every Idea organizes experience in such a way as to produce systematic and general knowledge that is expressed by means of natural laws. To get science, we need ideas that are clear and distinct, and that adequately colligate facts in such fashion that general propositions about matters of fact become possible. If, in turn, the general propositions (laws) support one another through what Whewell calls a "consilience of inductions,"[9] and provide the basis for the deductive derivation of further truths, one can see intuitively that these general propositions (or the axioms from which they are derivable) are not only true, but necessarily true: that is, their logical negations are incapable of clear and distinct conception, "even by an effort of imagination, or in a supposition."[10]

Throughout his writings on the subject, Whewell insists upon two fundamental features of necessary truth. First, no necessary truth is derivable from experience, and second, a necessary truth is one whose negation is not only false, but impossible—that is, "necessary truths are those of which we cannot distinctly conceive the contrary." These two propositions characterizing necessary truth together imply a third: necessary truths are not known by means of discursive reasoning, nor are they, in Kant's sense of the term, analytic; necessary truths are known by intuition. Each of these three propositions needs to be understood in some depth before Whewell's general position that facts are idealized (that truths of experience become necessary truths) becomes really clear.

When Whewell says that we can derive no necessary truth from experience, he seems at first glance to be saying something quite inconsistent, if he also holds that necessary truths emerge as necessary in the course of development of merely empirical sciences. The apparent inconsistency disappears, however, when we look at Whewell's arguments, instead of at his statements of the view. The word "derived," which Whewell uses everywhere, does not fully express his intended meaning. What he does mean is that the *evidence* for necessary propositions is never empirical; that we cannot arrive at necessary truths by the simple expedient of collecting and listing facts. Thus, for example, we might seem to be confirming a simple arithmetical proposition empirically by adding objects together to get a sum. But Whewell believes that as soon as we conceive the numbers—not the objects counted—distinctly, we see the sum, and no number of repeated trials of counting objects will alter the truth of the proposition. Indeed, "we cannot be said to make a trial, for we should not believe the apparent result of the trial if it were different." Similarly, in discussing the statical principle, "the pressure on the support is equal to the sum of the bodies supported," Whewell makes the following point:

> But in fact, not only are trials not necessary to the proof [of this proposition], but they do not strengthen it. Probably no one ever made a trial for the purpose of showing that the pressure upon the support is equal to the sum of the two weights. Certainly no person with clear mechanical conceptions ever wanted such a trial to convince him of the truth; or thought the truth clearer after the trial had been made. If to such a person, an experiment were shown which seemed to contradict the principle, his conclusion would be, not that the principle was doubtful, but that the apparatus was out of order. . . . We maintain, then, that this equality of mechanical action and reaction, is one of the principles which do not flow from, but regulate our experience. To this principle, the facts which we observe must conform; and we cannot help interpreting them in such a manner that they shall be exemplifications of the principle.[11]

The result of this sort of argument is clear. No experience, even a highly organized and regularized experience brought about by an experiment, can confirm a necessary truth, because no conceivable experience could disconfirm it. Every apparent disconfirmation will either be traced to a mistake in procedure or will be interpreted to fit the law. Whewell here anticipates an important point that has received much attention in recent philosophy of science. No proposition can be regarded as empirically significant if there is no empirical evidence that could possibly disconfirm it. Such a proposition is either meaningless or analytic. However, though Whewell anticipates this point, he would not subscribe to it in its more recent form.

Apparently, then, we cannot interpret Whewell's view, that in the course of time empirically-known propositions come to be apprehended as necessary, to mean that the logical character of a proposition changes in time. Logically necessary propositions cannot be constructed out of logically contingent ones (which is not the same as saying that contingent propositions have no logically necessary implications). Whewell's argument against such a construction is always the same: contingent propositions that we learn from experience may be general, but they are never necessary. And since necessary propositions *are* necessary, it follows that they cannot be derived from experience. This means, of course, that the kind of evidence that establishes a necessary truth (for Whewell, intuition of its self-evidence) is different from the kind of evidence (observation or experiment) that establishes an empirical proposition. Whewell believed that this position is not incompatible with holding that truths known empirically come to be known subsequently as necessary.

Whewell was aware that the term "experience" could be construed differently in the two sentences (1) "Necessary truths are not derived from experience," and (2) "Some truths known by experience are later known to be necessary." In the first sentence, "experience" means for Whewell "observation or experiment in the context of some clear and precise scientific theory"; experience in this sense is thus regularized and deliberately accumulated. He speaks of this form of experience as the only one to which the name "can properly be applied," and says that

> this experience is distinctive; it implies not only the faculty of perceiving, but special objects perceived; not merely the perception of something, but the perception that things are and occur in a certain manner to the exclusion of any other manner. It [is] contingent; it not only comes after the first exercise of perception, but it may fail to come at all.[12]

In the second sentence, "experience" means "sense experience" or simply "perception." It is in this second sense that one can say that we know necessary truths as matters of experience prior to apprehending them as necessary.

> If it be said that we cannot possess the ideas of pressure and mechanical action without the use of our senses, and that this is experience; it is sufficient to reply that the same may be said of the ideas of relations in space; and that thus Geometry depends upon experience in this sense, no less than Mechanics. But the distinction of necessary and empirical truth does not refer to experience *in this sense*, as I need not now stop to show.[13]

Though Whewell was convinced that there are these two separate and distinct meanings of the term "experience," he was not always careful to keep the two senses distinct in his writings. It may have been this lack of clarity in expression that confused some of his critics and gave rise to some of the difficulties in understanding his view of the apparent empirical origin of necessary truths. There are at least two places in his systematic writings, however, where Whewell does make the attempt to be as clear as possible about the two meanings of "experience." Whewell contends[14] that "Science begins with *Common* Observation of facts, in which we are not conscious of any peculiar discipline or habit of thought exercised in observing." He also refers to this form of observation ("experience" in the second sense) as "observation of the plainest and commonest appearances" which is the exercise of "the mere faculties of perception." Experience in this sense is perception of appearances and recurrences of appearances of the most familiar things. It is in this sense of experience that we first observe the positions of the planets at different times, that we first become aware of the most obvious facts about bodies in motion, and that we first observe familiar aspects of visible objects. It is this sense of "experience" that Whewell intended to convey in such sentences as "some truths known by experience are later known to be necessary," and "we learn by experience."[15] Thus construed, experiences are individual perceptions of facts, and the nature of such facts will vary from individual to individual. Some perceptions will be clear, some confused, some will be of this aspect of a thing, some of another aspect. Given the somewhat idiosyncratic character of such perceptions, they cannot by themselves count as evidence for any scientific generalization, nor for any proposition that is necessarily true.

If we are to have science, or what Whewell sometimes calls "speculative knowledge," it must be possible for us to have experience of the first type, namely controlled observation or experiment. In one place, Whewell gives what is perhaps his clearest expression of what he means by "experience" in this sense. He writes:

I here employ the term Experience in a more definite and limited sense than that which it possesses in common usage; for I restrict it to matters belonging to the domain of science. In such cases, the knowledge which we acquire, by means of experience, is of a clear and precise nature; and the passions and feelings and interests, which make the lessons of experience in practical matters so difficult to read aright, no longer disturb and confuse us.[16]

Whewell gives several examples of the kinds of propositions that we can know by means of such scientific experience. We know that animals that ruminate are cloven-hoofed, and that all the planets and their satellites revolve around the sun from west to east. Similarly, we know by such intentional observation that all meteoric stones contain chrome. In order to have scientific experience, Whewell believes that the scientist must introduce clear and precise conceptions that, when they enter into the formulation of hypotheses about the actual course of events, allow us either to classify or to predict accurately. It was Whewell's view that the most efficacious scientific conceptions are those that permit us to make measurements or that permit deductive moves from one hypothesis to another in a scientific system. Whether or not the conceptions introduced by the scientist are quantitative in character, they must at least permit the organization of data in such fashion that hypotheses that are either true or false of that data become possible. What distinguishes scientific experience from experience as mere perception is the fact that scientific experience, unlike mere perception, arises from the deliberate imposition on the data of a concept that will permit a hypothesis that is traced deductively to its consequences to be either confirmed or refuted by precise observations or measurements that are not possible until the concept is introduced.

The distinction that Whewell seems to have in mind is perhaps brought out by comparing the sentence (A) "I perceive *x* as red" and the sentence (B) "This *x* is red." Whewell would have to interpret sentence (A) as a report of what he calls a "mere preception." Given this interpretation, it would be true to say, "I learn by experience that a red appearance enters my visual field." In addition, I could perhaps characterize the experienced red as harsh or soft, hazy or clear, but I could not claim, at least on the basis of the experienced red alone, that the object seen *is* red. Sentence (B) is to be distinguished from sentence (A) because sentence (B) is an objective knowledge claim that can be determined to be either true or false. In order to make such a determination, however, I would have to introduce a relatively clear and precise conception of color, such as that given in the conception of colors as wave-lengths of visible light. On the basis of this conception, I should then be able to determine whether or not the object I see has a color that falls within the wave-length range between 6220 and 7700 angstroms. Thus, spectral analysis would make possible those "scientific experiences" that would determine whether or not the claim, "This $x$ is red," is true. The important point to bear in mind is that, though the results of precise scientific observation and measurement can provide evidence for scientific hypotheses, they can never completely establish these hypotheses as universal laws. Experiences as mere perceptions, on the other hand, though they may give rise to, or make us aware of states of affairs or propositions, never provide evidence for or against any objective knowledge claims. On the basis of this distinction, Whewell is able to hold both that necessary truths are learned by experience (in the sense of perception) and that these truths are not established as necessary by appeal to experience (in the sense of controlled scientific experience).

The two meanings of "experience" play fully operative roles in Whewell's theory. Given the meaning of "experience" as "scientific experience," Whewell was able to argue for his logical distinction between necessary and factual truths. Given the meaning of "experience" as "sensation" or "perception," he was able to hold that no knowledge is innate, and that thus even necessary truths take some time and experience to learn. Whewell also believed that, presupposing that no Ideas are ultimate, there are still more of them to be discovered in the future as new sciences become possible, and as subsequent discussion clarifies Ideas to such an extent that they become clear and distinct, and hence become sources of intuitively certain necessary propositions. It seems, then, that Whewell had formulated a logical basis for holding both that there is a logical distinction to be made between empirical and necessary propositions, and that some truths known empirically at first are subsequently known to be necessary.

In spite of his careful attempt to make a logical distinction between necessary and empirical statements, Whewell's position is somewhat puzzling, since it is clear that he also assumes that statements asserting necessary truths are nontrivial (nonanalytic), and in some sense factually meaningful. Indeed, they are to be construed as the very source of meaningfulness in scientific systems, and they could not play this role, at least not so far as Whewell is concerned, unless they had some sort of reference to ontological realities. I think it obvious that Whewell does not want us to regard necessary truths as analytic: that is, as true by virtue of being logically necessary. For his view that necessary truths are those whose negations are not clearly and distinctly conceivable is not translatable into the assertion that necessary truths are those from whose negations logical contradictions are derivable, even though this derivation might in fact be possible. In the course of his discussion with Mansel on the question of necessary truth, Whewell explicitly denies that the type of necessity characteristic of propositions that express the Fundamental Ideas is logical necessity. "I will not pretend to say that this kind of necessity is exactly represented by any of those Fundamental Ideas which are the basis of science. . . ."[17] Also to be taken into account is Whewell's continuing argument, first against Stewart and then against Mill, to the effect that both definitions and self-evident axioms are necessary as the basis of mathematical reasoning, and that mathematical reasoning is therefore not hypothetical, but has an ontological reference. In addition, Whewell argues at some length that not even basic arithmetical and geometrical propositions are necessarily true by virtue of their logical necessity. Without actually referring to propositions in arithmetic and geometry as synthetic a priori propositions, Whewell does produce arguments that closely parallel Kant's arguments for the synthetic apriority of such propositions, including the arguments that the objects (space and time) of arithmetic and geometry are intuitable.

For example, Whewell argues that it is not true that the assertion "3 plus 2 equals 5" "merely expresses what we mean by our words; that it is a matter of definition; that the proposition is an identical one" (a tautological one). Indeed, it is not even true that the definition of 5 is "3 plus 2." Rather, the definition of 5 is "4 plus 1." But how is it that 3 plus 2 is the same number as 4 plus 1?

Not because the proposition is identical; for if that were the reason, all numerical propositions must be evident for the same reason. If it be a matter of definition that 3 and 2 make 5, it must be a matter of definition that

> 39 and 27 make 66. But who will say that the definition of 66 is 39 and 27? . . . How do we know that the product of 13 and 17 is 4 less than the product of 15 and 15? We see that it is so, if we perform certain operations by the rules of arithmetic; but how do we know the truth of the rules of arithmetic?

The correctness of the rules can be rigorously demonstrated. Perform this operation, and such-and-such must inevitably be the result.

> Certainly this can be shown to be the case. And precisely because it *can* be shown that the result must be true, we have here an example of a necessary truth; and this truth, it appears, is not *therefore* necessary because it is evidently identical, however it may be possible to prove it by reducing it to evidently identical propositions.[18]

Whewell does not interpret the second major characteristic of necessary truths—that we cannot distinctly conceive their negations—to mean that necessity is simply logical necessity. Rather, the necessity of propositions should be traced back to the categorial necessity of the Fundamental Ideas that they express, and that form them. This brings us to the third essential feature of necessary truths, namely, that we know them by intuition of a certain kind. Whewell says

> the way in which those Ideas became the foundation of Science is, that when they are clearly and distinctly entertained in the mind, they give rise to inevitable convictions or intuitions, which may be expressed as *Axioms;* and these Axioms are the foundations of Sciences respective of each Idea.[19]

Whewell's critics misunderstood this view, because his expositions of it did not always state it clearly. Most especially, he does not mean that the Fundamental Ideas are innate possessions of every mind, nor does he mean that everyone in whatever condition of mental development will be able clearly and distinctly to apprehend necessary truths. Nor does the fact that some persons (children and mental defectives, for example) are unable to conceive clearly and distinctly the axioms of, say, geometry, mean that geometry thereby loses its universality and necessity. The attainment of a state of mind requisite for intuiting necessary truths is a cultural and educational development.

Actually, Whewell's critics need not have been misled. Throughout his writings he insisted that the Ideas must be "clearly and distinctly possessed" before they can become sources of intuitively necessary axioms. He also tried to make it clear that the intuition of necessary truths is a "rare and difficult attainment,"[20] coming only in that stage in the development of a science when the requisite categorical forms or Ideas have been simplified and organized (in short, rendered clear and distinct), and when properly trained and suitably ingenious scientists are available. For Whewell, this developmental thesis—"*There are scientific truths which are seen by intuition, but this intuition is progressive*"[21]—had a quite concrete historical meaning. No particular result of scientific investigation yields necessary truths. Rather, it is the existence, the *possibility,* of a science that establishes the necessity of its axiomatic principles.[22] Thus we know that particular Ideas are distinctly conceived when, in the course of actual history, they initiate inductive sciences that really do explain and predict the phenomena they were introduced to deal with. Ideas that in fact organize and systematize whole bodies of general propositions are really fundamental; and such organized and systematized bodies of general propositions are precisely what, for Whewell, are to be counted as sciences.

More particularly, we know that a man has distinct conceptions if he can comprehend the axioms and follow the reasoning in any science, either formal or empirical. Though this proposition means that there are necessary truths that are not so known to everyone, Whewell is not bothered by the apparent implication that certain persons who are capable of clear conception, in effect create truths that depend for their necessity on these persons' ability to think clearly and distinctly. For when a man has knowledge,

> We conceive that he knows it because it is true, not that it is true because he knows it. . . . We are not surprised that attention and care and repeated thought should be requisite to the clear apprehension of truth. For such care and such repetition are requisite to the distinctness and clearness of our ideas: and yet the relations of these ideas, and their consequences, are not produced by the efforts of attention or repetition which we exert. They are in themselves something which we may discover, but cannot make or change.[23]

It follows from this proposition that all axioms are, in Kant's sense of the term, "constitutive" principles, or those principles without which knowledge would be impossible. Logic does not constrain us to think in the terms provided by the axioms—we would not be guilty of simple logical contradiction if we entertained the concept of an uncaused event, for example—but our desire to have knowledge of reality does so constrain us. For surely we cannot know except by means of those principles that make knowing possible.

## WHEWELL'S THEORY OF INDUCTION

Recent discussion of the so-called deductive model of scientific explanation has drawn attention to certain features of hierarchical or subsumption types of scientific systems, in which statements in the observation language are subsumed under (explained by) empirical generalizations, which in turn are subsumed under (explained by) higher level theoretical generalizations. Whewell's theory of science as successive generalization is one early form of this kind of philosophy of science. His theory of induction, in its full form, expresses what is now called the hypothetico-deductive character of well-developed sciences.[24]

In his *Philosophy of the Inductive Sciences* Whewell puts forward the view that science progresses from empirical generalizations of observed particulars to more and more comprehensive and inclusive theoretical generalizations. This now-familiar and frequently-noted progression establishes—in the order of discovery—a ladder of generalizations ranging from low order empirical generalizations to very inclusive generalizations like Newton's Gravitational Law. Inverting the ladder yields the order of verification, since each of the lower order generalizations is both derivable deductively from the one of next highest order, and is the inductive evidence for that higher order generalization. Whewell's unique contention was that such an inductive ladder could be written down in the form of what he called an Inductive Table,[25] in which the subsumed generalization would be marked off from the subsuming generalization by a kind of bracket, much as the conclusion of a syllogism can be marked off from its premises by a line. Thus, for example, Whewell's proposed inductive table for astronomy demonstrates that the science takes its beginning from such simple observed facts as "The earth appears to be immovable," "The moon's bright part is of the shape of a ball enlighted by the sun," "The tides ebb and flow," and other fairly common and obvious observations. These ordinary facts are then generalized through successive inductive steps, proceeding historically from the Greeks through Copernicus and Kepler to Newton's Theory of Universal Gravitation. Such compact tabular arrangements (Whewell also refers to them as "genealogical trees") exhibit the logical structure of induction. Whewell believed that one could see by inspection that the facts and the inference from them were put down in such a manner that the evidence supported the inference made by induction. In effect, the entire table is a set of inductive formulae, and Whewell thought that such formulae corresponded roughly to the syllogistic formulae available in deductive logic. Before probing more deeply into Whewell's theory of the logic of induction, it would be desirable to point out some obvious features of these tabular arrangements of scientific hypotheses.

First, Whewell thought that the tables clearly exhibited two of the most prominent features of progress in science: the tendency to simplicity that is observable in all true scientific theories, and what Whewell called "the consilience of inductions." According to Whewell, "*The Consilience of Inductions* takes place when an Induction, obtained from one class of facts, coincides with an Induction, obtained from another different class."[26] He thought that a consilience of inductions was involved when Newton used Kepler's laws to explain the central force of the sun, which is the universal force of all heavenly bodies, and the precession of the equinoxes. As we will see below, consiliences take place when different classes of facts are explained by the same hypothesis, and when predicted facts are unexpected or novel. In this respect consilience has to do with what is now called independent evidence for a hypothesis.

The tendency to simplicity has direct reference to what might be called the power of inclusiveness of a hypothesis, or what Whewell would have been fonder of calling "the colligating power" of the concepts involved in a generalization. I will not pause here over the details of Whewell's view of simplicity, but it should be noted that he appears to be one of the first to construe the simplicity of scientific theories precisely, not in vague aesthetic terms. One can show in Whewell's writings that by the simplicity of a concept or the simplicity of a hypothesis, he had in mind an extensionalistic notion of simplicity. One concept is simpler than another if it colligates more facts, if it has a larger denotation. It is easier to construe simplicity in these extensionalistic terms when one is dealing with metrical concepts. Perhaps this is why Whewell held that the third step in all inductions was the "determination of the magnitudes"; i.e., the translation of the concept into mathematical terms. However this may be, it surely is the case that Newton's concept of universal gravitation has a larger extension and hence, at least in this sense, is simpler than Kepler's concept that the curves described by planets are ellipses.[27]

Though Whewell thought this now-familiar "Chinese box" or subsumption view of scientific theories did illustrate some important features of induction, i.e., the consilience of inductions and the tendency of scientific theories to become simpler and simpler as successive inductive generalizations take place, he did not think that the inductive tables represented the entire process of induction in each case. Especially noteworthy is the fact that the inductive tables that can be written down for well-developed sciences give the appearance that science develops by mere collection of particulars and in turn by collection of these generalizations, in higher level generalizations. Whewell insisted throughout his writings on induction, and particularly in his well-known controversy with Mill, that induction is not a mere description of phenomena, nor a mere collecting of data. Rather, the critical aspect of an inductive process is what Whewell called the introduction of a new conception, a "principle of connexion and unity, supplied by the mind, and superinduced upon the particulars." Thus, Kepler's law that the planets describe elliptical orbits is for Whewell neither a collection of observations nor a simple description of what is apparent by observation and calculation. It involves a deliberate imposition upon the data of the concept of an ellipse, without which the induction could not have been made in the first place. As Whewell puts it in one place, "The Inductive truth is never the mere *sum* of the facts. It is made into something more by the introduction of a new mental element; and the mind, in order to be able to supply this element, must have peculiar endowments and discipline."[28] Thus, though each inductive inference must be carefully verified by comparing it with the particular facts, this process cannot happen except for those with properly disciplined, trained use of scientific conceptions. Whewell insists that the discovery of scientific truth requires both scrupulous and painstaking attention to the data and equally rigorous discussion and clarification of our ideas.

This is perhaps a peculiar view of the kind of inferring that takes place in an induction, but there is much merit in the suggestion that though particular facts are included in an inductive generalization, they are not merely included. We might learn something from a simple enumeration of the facts; indeed, we might even display in this enumeration the evidence for some inductive move. But surely if one views induction as Whewell did, as a kind of process of discovery rather than as a mode of proof or of argument, then Whewell was right when he claimed that "the inductive step consists in the *suggestion* of a conception not before apparent."[29]

Without understanding Whewell's special sense of induction as the colligation of several facts in a single proposition by the introduction of a new concept, one cannot make any sense at all out of some of the strange claims that he makes for the inductive tables. Indeed, unless Whewell's meaning for induction is kept clearly in mind, one also cannot understand the contention that the inductive tables display the *logic* of induction; for Whewell, the logic of induction is the logic of discovery, as much as the logic of proof. Contrary to Mill's thinking, Whewell did indeed discuss the logic of proof and his discussion ranges over the usually noted tests of hypotheses, details of measurement, and methods of calculation.[30] But, unlike most contemporary methodologists, Whewell regarded inductive discoveries as having a logic of their own. He said, "The analysis of doctrines inductively obtained, into their constituent facts, and the arrangement of them in such a form that the conclusiveness of the induction may be distinctly seen, may be termed *the Logic of Induction*".[31] By logic in this context Whewell means "a system which teaches us so to arrange our reasonings that their truth or falsehood shall be evident in their form."

There is a more than apparent strangeness about the concept of logic involved in Whewell's discussion of the inductive tables. It would have been more in keeping with views of logic current in the middle of the nineteenth century for Whewell to say that the logic of induction consisted in the laws or rules by means of which we obtain proper inductions from either particulars or low order generalizations. However, even in syllogistic deductive reasoning, it was not Whewell's view that the syllogism provides us with rules by means of which valid inference can be carried on. Instead, he seems to think of logic as consisting simply in a compendious and compact schematic presentation of what had already been thought in more complex manners. In addition, with regard to inductive reasoning in science, Whewell believed that, given the imposition of a new concept on a set of data, the scientist must try this concept in hypothetical form to determine whether the data really would support it, to determine whether this concept might be the one that would colligate these data, bringing them together in a manner that would be both useful and true. The procedure here is one of trial and error, and in many cases it involves the work of numerous scientists investigating over long periods of time. Hence, the logic of induction is simply the schematization of both the introduced concept and the laborious procedure of testing the concept against the data.

What the inductive table makes possible, then, is not a rigorous derivation of high-order generalizations from the observation statements that they ultimately subsume, but rather what Whewell refers to as a special *act of attention*.[32] The inductive table, in other words, helps us to see what, for Whewell, are real connections among the data, once they are properly colligated by appropriate and clear conceptions. As he puts it, "The scheme being clearly conceived, we *see* that all the particular facts *are* faithfully represented by it; and this agreement, along with the simplicity of the scheme, in which respect it is so far superior to any other conception, . . . persuade us that it is really the plan of nature."[33] In this respect then, the inductive table enables us to intuit the rightness of a connection between a certain concept and its data, or between two or more concepts. But for Whewell this intuitive character of right induction is also shared by proper deduction, because no matter how long the chain of demonstrations might be, they all lead eventually to a perception of a connection between the data or between the premises and a conclusion. Whewell's view of deduction, then, is very much like that of Descartes. However long the chain of reasoning might be, we must see at each step in that reasoning the correctness of the appropriate move. But Whewell adds something to this Cartesian concept of deduction in viewing the difference between deduction and induction as merely a difference in direction. He writes:

> Deduction is a necessary part of Induction. Deduction justifies by calculation what Induction had happily guessed. Induction recognizes the ore of truth by its weight; Deduction confirms the recognition by chemical analysis. Every step of Induction must be confirmed by rigorous deductive reasoning, followed into such detail as the nature and complexity of the relations (whether of quantity or any other) render requisite. If not so justified by the supposed discoverer, it is *not* Induction.[34]

The point may seem obvious. If one is to hold a subsumption theory of science, then there must be some way of guaranteeing that lower-order generalizations, and eventually observation sentences, really are included—in the sense of deductively included—in higher-level generalizations, and it is this sense of precise logical inclusion that Whewell believes the inductive tables must exhibit clearly.

This situation is puzzling because Whewell, as a mathematician and methodologist of science, must have realized that we cannot guarantee any such deductive sense of inclusion as determining the truth of an induction, unless the basic observation sentences of a given science contain predicates that take as their referents a finite class of particulars, every one of which has been inspected. Whewell's inductive tables, then, would seem to be adequate logical schematizations only of complete or perfect inductions; that is, inductions in which the general conclusion follows from a finite, completely inspected group of particulars. Unhappily, Whewell says nothing in detail about this technical problem in logic, and I think here are two related reasons for this omission. First, Whewell had in mind a basically Platonic view of the nature of scientific reasoning. He thought that scientific investigation clarified concepts rather than proved them in an inductive sense. This Platonism seems to me to be clear as an implication of what I have just been saying about

Whewell's view of intuitive connections in both induction and deduction. Indeed, he even holds in another work that experiments do not in any ordinary sense prove or confirm scientific laws, but rather play the role of rendering laws more clear and intelligible "as visible diagrams in geometry serve to illustrate geometrical truths."[35] Second, and this is perhaps the more fundamental of the two reasons for the puzzles that arise about Whewell's view of induction, he seems to have confused psychological conviction with objective empirical truth, and with inferential validity.

This psychologistic character of Whewell's thought is brought out by considering further his view that, in some cases at least, induction allows the properly disciplined scientific mind to see a necessary connection between the facts and the general proposition in which they are included. Thus Whewell holds that in cases of perfect induction "the conviction of the sound inductive reasoner does reach"[36] the point where he sees not only that the results can be explained by and hence deduced from a certain general proposition, but also that they can be explained by and deduced from only this general proposition and no other. Thus the ultimate formula for valid and perfect induction would be " 'The several Facts are exactly expressed as one Fact if, *and only if*, we adopt the Conception and the Assertion' of the inductive inference."[37] Though Whewell believed that there exist such cases of completely accurate intuition of the necessity of an inductive inference, he holds that the conviction "that the inductive inference is not only consistent with the facts, but necessary, finds its place in the mind gradually, as the contemplation of the consequences of the proposition, and the various relations of the facts becomes steady and familiar."[38] This view of course is consistent with Whewell's general point that the progress of science consists in a movement toward more and more inclusive propositions and toward propositions that scientists will eventually see as self-evident truths. Whewell puts this point in various ways, but the following quote is a representative statement.

And thus we see how well the form which science ultimately assumes is adapted to simplify knowledge. The definitions which are adopted, and the terms which become current in precise senses, produce a complete harmony between the matter and the form of our knowledge; so that truths which were at first unexpected and recondite become familiar phrases, and after a few generations sound, even to common ears, like identical propositions . . . [All sciences illustrate] . . . the general transformation of our views from vague to definite, from complex to simple, from unexpected discoveries to self-evident truths, from seeming contradictions to identical propositions.[39]

The history of science consists in the record of a long and laborious chain of trial and error investigations. The conceptions that the scientists introduced in an attempt to colligate the facts available at any particular period of time are tested in this procedure. Whewell thought that any particular science would eventually arrive at generalizations so inclusive and so simple that they would be seen as propositions that not only expressed the facts but expressed the facts in a way that the mind would apprehend as necessary. Thus science does not begin with necessary truths, but arrives at them, and the necessity is ultimately guaranteed by intuition.[40]

Because Whewell held that the development of the natural sciences would eventually arrive at the apprehension of necessary truths, he could also hold, as the last point about the inductive tables, that such tabular arrangements of general propositions may be considered as the "*Criterion of Truth* for the doctrines which they include."[41] The tables may be viewed as such a criterion because each of the general propositions that is written down truly subsumes certain other generalizations, and ultimately observation sentences. Thus a high-level generalization in science collects together, i.e. colligates, already ascertained truths. Ultimately, then, a very high-level scientific generalization collects together or asserts a great body of facts, as Whewell puts it, "duely classified and subordinated." On this point, also, Whewell is entitled to hold that the inductive tables function as a criterion of truth, given his motion that at least some inductions yield necessary conclusions. To say that the inductive tables function as a criterion of truch is again equivalent to saying that they schematize what is intuitable as necessary.

I have outlined a reading of Whewell's theory of science that shows how he can be thought to have tied together his views on necessary truth and on demonstrative induction. This reading emphasizes Whewell's platonism and his apparent psychologism. In stressing that colligations of facts eventually become *the* way to view them, Whewell was arguing, in effect, that laws of nature turn out to be also laws of thought. After all, if one is somehow constrained to see the facts in this one way and in no other, if in all cases of proper induction only one interpretation is finally possible, then surely we have discovered not only a truth about nature, but a truth about what it is impossible for us to think otherwise. In Whewell's day it was a widespread belief that the laws of deductive logic are psychological laws, laws of thought.[42] It is clear that there is a logic of deduction; if induction also has a logic, as Whewell thought that it did, then it would be natural for him to presuppose that something psychologically lawlike is involved in the process of inductive inference. As we have seen, this psychological component is the "act of attention" made possible by the inductive tables. What the tables enable us to do is to "see" that the facts could not have been correctly expressed in any other way. We grasp the necessity of the law or theory.

The psychologistic reading of Whewell thus has both textual support and fits neatly into nineteenth-century British modes of thought about logic.[43] This reading is also dominant in the accepting reaction to Whewell that we find in Ernst Mach's *Knowledge and Error* (1905). Earlier, Karl Pearson's *The Grammar of Science* (1892), had argued for the importance of disciplined imagination in the formation of scientific hypotheses. Both Mach and Pearson stress that the aim of science is the production of a certain "economy of thought" that results from our attempts to unify our knowledge and to arrive at an explanation of the facts that is the most simple we can achieve. Mach had Whewell's model of inductive inference explicitly in mind when he formulated his views on scientific method. I think Pearson did also, although he never acknowledged any debt to Whewell. The point is that there are influential theories of the development of science that trade on Whewell's idea that inference in science involves the step that is beyond the reach of method. Perhaps, though, the similarity ends there. Whewell insisted that the laws of nature we discover by the application of the inductive method are *truths* about nature, and *necessary* truths in the bargain. Both Mach and Pearson (and later positivistic followers) thought that the laws of nature are *just* economies of thought, compendious shorthand expressions allowing us the luxury of not having to deal with all of the facts, one by one. The generality of a scientific law has to do with its capacity to collect a large number of particulars under one rubric; 'all' means only a conjunction of particulars (a view also argued for by John Stuart Mill).

Pearson regarded the maxim that we should seek the most simple ways of expressing nature's regularities as psychological advice. It was Schlick, in his *General Theory of Knowledge* (1918), who pointed out that seeking descriptions of natural regularities that require "the least expenditure of thought" is an enterprise allowing two interpretations. It can be taken to be a factual claim about the nature of science, in which case it is false. Far from being the psychologically most simple way of describing nature, science requires enormous expenditures of thought. Most people do not find mathematics easy to learn and to use. But we can also read the maxim of economy of thought as a kind of epistemological advice, understanding "least expenditure of thought" in the *logical* sense, where what we mean is that scientists should seek logical simplicity in their theories. This is equivalent to holding that scientists should seek expressively very rich concepts for describing nature, that theories which say more using few concepts are to be preferred to those saying less with a greater number of concepts.

Schlick's analysis of scientific systems and of patterns of scientific inference directs our attention to the logical, not the psychological, features of scientific theories. If we follow this lead, a second interpretation of Whewell becomes possible, one that endeavors to embed his thoughts about apparently psychological matters in a context of logic and epistemology in conformity with more recent ways of thought in philosophy of science. It appears unlikely that he would have accepted any form of instrumentalism or antirealism as a correct theory of science. Despite the marked psychologistic features of scientific inference his theory of science emphasized, his philosophical instincts are all on the side of realism. Science is the right interpretation of nature, not merely the simplest way to talk about it.

On this second reading we are to emphasize that Whewell's logic of induction deals with various levels of *test* of theories or hypotheses, indeed, of colligations, since each superimposed idea counts as a theory about some class of data we wish to explain. We are invited to think of Whewell's logic of induction as a theory of confirmation. Discussion of this second version of Whewell's theory of induction will help to highlight different points of emphasis and will shed more light on what Whewell was trying to accomplish. The second interpretation, like the first, has textual support in Whewell's writings, and also fits into other nineteenth-century modes of thought, for example, Sir John Herschel's work on probable inference.

I have pointed out above that for Whewell a colligation of the facts involves the superimposition of a new idea on some set of data. Induction begins with this mental act. The content of that act, however, is a proposition that claims to be true of the data, or facts. Hence a colligation in its initial form either fits or fails to fit a given body of data. Its evidential strength comes from its power to illuminate a single kind of object or event. In time, scientists discovered that certain kinds of ideas are more successful as ingredients in colligating acts than are others. For example, ideas that can be quantified have been more successful in providing good inductions than have ideas that cannot. A proper scientific interpretation of some class of data is therefore already a highly theoretical one. If the data must be reducible to quantitative treatment, then a colligation is something like a Newtonian "phenomenon"; in Whewell's terms, it is "more than the sum of the facts," it is a way of seeing the facts, interpreting the facts, in a "new light."

However, a successful colligation may be self-stultifying with respect to its evidence, and it may have equally successful competitors. What in the way of additional independent evidence for a colligation is needed? A colligation, remember, explains only what has already been observed. All that is required of it is that it be adequate to the task of explaining what has been observed up until now. For Whewell, and for many other theorists about induction, it is important to note that a good colligation is viewed as a stronger explanation, a better-evidenced explanation, if it successfully predicts phenomena not yet observed.[44] This criterion of additional evidential support applies only to those cases where an explanation yields successful predictions of events of the same kind. In other words, the original colligating idea is reinforced by a further accumulation of predicted observations that turn out to be true. The sentences describing these observations are all deductive entailments of the hypothesis, theory or law.

Although it is an improvement over simple colligation, successful prediction is not the highest level of test of a theory. Whewell thought that the best test of any scientific explanation or theory is what he called *consilience.* A consilience of inductions explains data of a kind different from those it was initially introduced to explain. His claims for the independent evidential strength of a consilience may seem extravagant:

> The instances in which this [consilience] has occurred, impress us with a conviction that the truth of our hypothesis is certain. No accident could give rise to such an extraordinary coincidence. No false supposition could, after being adjusted to one class of phenomena, exactly represent a different class, where the agreement was unforeseen and uncontemplated. That rules springing from remote and unconnected quarters should thus leap to the same point, can only arise from *that* being the point where truth resides.[45]

One form of consilience is successful prediction of that which is unexpected or surprising. Every properly formulated theory has logical entailments as empirical claims following deductively from the theory. But if a theory successfully explains descriptions of kinds of things or events that are not mere logical consequences of a theory, but that provide additional independent evidence for the theory, we seem to be involved in some kind of reduction of classes of data to other kinds of classes of data. This is indeed what Whewell proposes: A proper consilience of inductions takes place when data of a kind different from the deductive expectations of a theory are seen to be of the *same kind* as those initially colligated. Clearly, this can take place only when the meanings of terms are changed to accommodate inclusion of the lower level hypotheses or laws under the consiliating theory. Colligating ideas preferred at earlier times are later replaced or altered by the semantics of the more general theory or law. If this did not happen, the deductive inclusion of the earlier hypotheses could not be guaranteed.

This is not the place to offer a reconstruction of Whewell's seminal idea of consilience in the format provided by present-day probability theory and alternative theories of confirmation.[46] It must be pointed out, however, that a probabilistic theory of confirmation will show that consiliences of the kind Whewell had in mind do indeed raise the probability of the initial hypotheses. For the purposes of this introductory account of Whewell's theory of induction, it is more important to note that he is talking about increase in evidential strength of theories of enormous unity or simplicity. Successful explanation in science must now be thought of as theoretical unification. Just look again at the inductive tables of astronomy and optics inserted in Chapter III. What is being claimed for Newton's theory of universal gravitation is that it consiliates all of previously successful astronomy and in fact shows that the initial thought-to-be disjoint classes of data are really items of the same kind. The apparently immovable earth, the tides, the motions of the stars, the behavior of the sun and moon are all of the same kind, in the precise sense that the regularities involved are all governed by the inverse-square law. What it is about these data that is truly knowable as subject to law is that they are all gravitational phenomena. The consilience of inductions thus takes place when the theory of universal gravitation effectively reinterprets classes of data originally thought to be different in kind as now being the same in kind. Consilience is in this sense a license for changing the semantics of earlier theories.

It is now possible to give a general characterization of the kinds of theories Whewell thought to be most successful, indeed, thought to be incontrovertibly true. Consilience is a property of those theories having large deductive content. The predicates (Whewell's "ideas") employed in such theories will be few in number and will be expressively very rich. This means that the theories are so general that they have reached some point of unity, which for Whewell is the same as saying that they are simple. Obviously, consilient theories must provide the *best* explanation of the phenomena involved. Finally, and this seems to me to reflect Whewell's complete confidence in a realist interpretation of science, such theories must have arrived at that historical point where further testing is taken to be irrelevant to acceptance of the theories. Instead, any apparently negative results will be construed as calls for refinement, rather than as disconfirmations.

This last characteristic of fully consilient theories again exhibits Whewell's insistence that science yields necessary truths. As a theory becomes successively more general, acquiring the features of simplicity and unity as well, it comes to be thought of as a *completed* science. Its laws are taken to be in no need of additional evidence; they are necessary truths. Whewell certainly thought this to be true of Newton's theory of universal gravitation (and thus thought of his own work on the tides as simply extending and refining Newton's theory) and also of Newton's mechanics. [See Whewell's essay "On the Nature of the Truth of the Laws of Motion," Chapter II below.]

Perhaps Whewell's insistence upon the idea that scientific laws are necessary truths resulted from his Victorian optimism about the progress of science, a progress he took to be theoretically without limit but constrained by continuing developments in the essentially deductive enlargements of existing theories. The inductive tables graphically display this kind of progress. Moreover, Victorian optimism about science and its engineering applications was easily reinforced by the very success of the Newtonian synthesis and its new and marvelous applications in industry and other areas. Later in the era J. J. Thompson would still find it appropriate to make the Whewellian remark that the future of physics lies in the seventh decimal place. Refinement and extension of existing theories was the order of the day; complete overthrow of an established science like mathematical physics was unthinkable.

I will not here debate the merits of the two interpretations of Whewell I have outlined. Historians might prefer the platonic Whewell; philosophers, the Whewell who tries to account for different levels of evidence. The two readings have one thing in common: they both show that for Whewell science is about getting at the truth. Induction viewed as enhanced illumination of ideas and induction viewed as the logic of evidence are both of them induction understood as a methodical way of getting theories to participate in what Whewell calls "the surrounding medium of truth."[47]

Chapter *i*

# *Whewell's Early Theory of Induction and Its Philosophical Basis*

# REMARKS ON MATHEMATICAL REASONING AND ON THE LOGIC OF INDUCTION*

***Sect. I. On the Grounds of Mathematical Reasoning***

. . . . . . . . . . . . . .

6. We shall . . . make some remarks on the nature and principles of reasoning, especially as far as they are illustrated by the mathematical sciences.

Some of the leading principles which bear upon this subject are brought into view by the consideration of the question, "What is the foundation of the certainty arising from mathematical demonstration?" and in this question it is implied that mathematical demonstration is recognised as a kind of reasoning possessing a peculiar character and evidence, which make it a definite and instructive subject of consideration.

7. Perhaps the most obvious answer to the question respecting the conclusiveness of mathematical demonstration is this;—that the certainty of such demonstration arises from its being founded upon *Axioms;* and conducted by steps, of which each might, if required, be stated as a rigorous *Syllogism.*

This answer might give rise to the further questions, What is the foundation of the conclusiveness of a Syllogism? and, What is the foundation of the certainty of an Axiom? And if we suppose the former enquiry to be left to Logic, as being the subject of that science, the latter question still remains to be considered. We may also remark upon this answer, that mathematical demonstration appears to depend upon Definitions, at least as much as upon Axioms. And thus we are led to these questions:—Whether mathematical

*From *The Mechanical Euclid, containing the Elements of Mechanics and Hydrostatics demonstrated after the manner of The Elements of Geometry* (Cambridge and London 1837), 143–82.

demonstration is founded upon Definitions, or upon Axioms, or upon both? and, What is the real nature of Definitions and of Axioms?

8. The question, What is the foundation of mathematical demonstration? was discussed at considerable length by Dugald Stewart;[1] and the opinion at which he arrived was, that the certainty of mathematical reasoning arises from its depending upon *definitions.* He expresses this further, by declaring that mathematical truth is hypothetical, and must be understood as asserting only, that *if* the definitions are assumed, the conclusion follows. The same opinion has, I think, prevailed widely among other modern speculators on the same subject, especially among mathematicians themselves.

9. In opposition to this opinion, I urge, in the first place, that no one has yet been able to construct a system of mathematical truth by means of definitions alone, to the exclusion of axioms; although attempts having this tendency have been made constantly and earnestly. It is, for instance, well known to most readers, that many mathematicians have endeavoured to get rid of Euclid's "Axioms" respecting straight lines and parallel lines; but that none of these essays has been generally considered satisfactory. If these axioms could be superseded, by definition or otherwise, it was conceived that the whole structure of Elementary Geometry would rest merely upon definitions; and it was held by those who made such essays, that this would render the science more pure, simple, and homogeneous. If these attempts had succeeded, Stewart's doctrine might have required a further consideration; but it appears strange to assert that Geometry is supported by definitions, and not by axioms, when she cannot stir four steps without resting her foot upon an axiom.

10. But let us consider further the nature of these attempts to supersede the axioms above mentioned. They have usually consisted in endeavours so to frame the definitions, that these might hold the place which the axioms hold in Euclid's reasoning. Thus the axiom, that "two straight lines cannot enclose a space," would be superfluous, if we were to take the following definition:—"A line is said to be *straight,* when two such lines cannot coincide in *two* points without coinciding *altogether.*"

But when such a method of treating the subject is proposed, we are unavoidably led to ask,—whether it is allowable to lay down such a definition. It cannot be maintained that we may propound any form of words whatever as a definition, without any consideration whether or not it suggests to the mind any intelligible or possible conception. What would be said, for instance, if we were to state the following as a definition, "A line is said to be *straight* (or any other term) when two such lines cannot coincide in *one* point without coinciding altogether?" It would inevitably be remarked, that no such lines exist; or that such a property of lines cannot hold good without other conditions than those which this definition expresses; or, more generally, that the definition does not correspond to any conception which we can call up in our minds, and therefore can be of no use in our reasonings. And thus it would appear, that a definition, to be admissible, must necessarily refer to and agree with some conception which we can distinctly frame in our thoughts.

11. This is obvious, also, by considering that the definition of a straight line could not be of any use, except we were entitled to apply it in the cases to which our geometrical propositions refer. No definition of straight lines could be employed in Geometry, unless it were in some way certain that the lines so defined are those by which angles are contained, those by which triangles are bounded, those of which parallelism may be predicated, and the like.

12. The same necessity for some general conception of such lines accompanying the definition, is implied in the terms of the definition above suggested. For what is there meant by "*such* lines?" Apparently, lines having some general character in which the property is necessarily involved. But how does it appear that lines may have such a character? And if it be self-evident that there may be such lines, this evidence is a necessary condition of this (or any equivalent) definition. And since this self-evident truth is the ground on which the course of reasoning must proceed, the simple and obvious method is, to state the property *as* a self-evident truth; that is, as an axiom. Similar remarks would apply to the other axiom above mentioned; and to any others which could be proposed on any subject of rigorous demonstration.

13. If it be conceded that such a conception accompanying the definition is necessary to justify it, we shall have made a step in our investigation of the grounds of mathematical evidence. But such an

admission does not appear to be commonly contemplated by those who maintain that the conclusiveness of mathematical proof results from its depending on definitions. They generally appear to understand their tenet as if it implied *arbitrary* definitions. And something like this seems to be held by Stewart, when he says that mathematical truths are true *hypothetically*. For we understand by an hypothesis a supposition, not only which we may make, but may abstain from making, or may replace by a different supposition.

14. That the fundamental conceptions of Geometry are not arbitrary definitions, or selected hypotheses, will, I think, be clear to any one who reasons geometrically at all. It is impossible to follow the steps of any single proposition of Geometry without conceiving a straight line and its properties, whether or not such a line be defined, and whether or not its properties be stated. That a straight line should be distinguished from all other lines, and that the axiom respecting it should be seen to be true, are circumstances indispensable to any clear thought on the subject of lines. Nor would it be possible to frame any coherent scheme of Geometry in which straight lines should be excluded, or their properties changed. Any one who should make the attempt, would betray, in his first propositions, to all men who can reason geometrically, a reference to straight lines.

15. If, therefore, we say that Geometry depends on definitions, we must add, that they are *necessary*, not arbitrary, definitions,—such definitions as we must have in our minds, so far as we have elements of reasoning at all. And the elementary hypotheses of geometry, if they are to be so termed, are not hypotheses which are requisite to enable us to reach this or that conclusion; but hypotheses which are requisite for *any* exercise of our thoughts on such subjects.

16. Before I notice the bearing of this remark on the question of the necessity of axioms, I may observe that Stewart's disposition to consider definitions, and not axioms, as the true foundation of Geometry, appears to have resulted, in part, from an arbitrary selection of certain axioms, as specimens of all. He takes, as his examples, the axioms, "that if equals be added to equals the wholes are equal," that "the whole is greater than its part;" and the like. If he had, instead of these, considered the more properly geometrical axioms,—such as those which I have mentioned; "that two straight lines cannot

enclose a space;" or any of the axioms which have been made the basis of the doctrine of parallels; for instance, Playfair's axiom, "that two straight lines which intersect each other cannot both of them be parallel to a third straight line;"—it would have been impossible for him to have considered axioms as holding a different place from definitions in geometrical reasoning. For the properties of triangles are proved from the axiom respecting straight lines, as distinctly and directly, as the properties of angles are proved from the definition of a right angle. Of the many attempts made to prove the doctrine of parallels, almost all professedly, all really, assume some axiom or axioms which are the basis of the reasoning.

17. It is therefore very surprising that Stewart should so exclusively have fixed his attention upon the more general axioms, as to assert, following Locke, "that from [mathematical] axioms it is not possible for human ingenuity to draw a single inference";[2] and even to make this the ground of a contrast between geometrical axioms and definitions. The slightest examination of any treatise of Geometry might have shown him that there is no sense in which this can be asserted of axioms, in which it is not equally true of definitions; or rather, that while Euclid's definition of a straight line leads to no truth whatever, his axiom respecting straight lines is the foundation of the whole of Geometry; and that, though we can draw some inferences from the definition of parallel straight lines, we strive in vain to complete the geometrical doctrine of such lines, without assuming some axiom which enables us to prove the converse of our first propositions. Thus, that which Stewart proposes as the distinctive character of axioms, fails altogether; and with it, as I conceive, the whole of his doctrine respecting mathematical evidence.

18. That Geometry (and other sciences when treated in a method equally rigorous) depends upon axioms as well as definitions, is supposed by the form in which it is commonly presented. And after what we have said, we shall assume this form to be a just representation of the real foundations of such sciences, till we can find a tenable distinction between axioms and definitions, in their nature, and in their use; and till we have before us a satisfactory system of Geometry without axioms. And this system, we may remark, ought to include the Higher as well as the Elementary Geometry, before it can be

held to prove that axioms are needless; for it will hardly be maintained, that the properties of circles depend upon definitions and hypotheses only, while those of ellipses require some additional foundation; or that the comparison of curve lines requires axioms, while the relations of straight lines are independent of such principles.

19. Having then, I trust, cleared away the assertion, that mathematical reasoning rests ultimately upon definitions only, and that this is the ground of its peculiar cogency, I have to examine the real evidence of the truth of such axioms as are employed in the exact Mathematical Sciences. And we are, I think, already brought within view of the answer to this question. For if the definitions of Mathematics are not arbitrary, but necessary, and must, in order to be applicable in reasoning, be accompanied by a conception of the mind through which this necessity is seen; it is clear that this apprehension of the necessity of the properties which we contemplate, is really the ground of our reasonings and the source of their irresistible evidence. And where we clearly apprehend such necessary relations, it can make no difference whatever in the nature of our reasoning, whether we express them by means of definitions or of axioms. We define a straight line vaguely;—that it is that line which lies evenly between two points: but we forthwith remedy this vagueness, by the axiom respecting straight lines: and thus we express our conception of a straight line, so far as is necessary for reasoning upon it. We might, in like manner, begin by defining a right angle to be the angle made by a line which stands evenly between the two portions of another line; and we might add an axiom, that all right angles are equal. Instead of this, we define a right angle to be that which a line makes with another when the two angles on the two sides of it are equal. But in all these cases, we express our conception of a necessary relation of lines; and whether this be done in the form of definitions or axioms, is a matter of no importance.

20. But it may be asked, If it be thus unimportant whether we state our fundamental principles as axioms or definitions, why not reduce them all to definitions, and thus give to our system that aspect of independence which many would admire, and with which none need be displeased? And to this we answer, that if such a mode of treating the subject were attempted, our definitions would be so

complex, and so obviously dependent on something not expressed, that they would be admired by none. We should have to put into each definition, as conditions, all the axioms which refer to the things defined. For instance, who would think it a gain to escape the difficulties of the doctrine of parallels by such a definition as this: "Parallel straight lines are those which being produced indefinitely both ways do not meet; and which are such that if a straight line intersects one of them it must somewhere meet the other?" And in other cases, the accumulation of necessary properties would be still more cumbersome and more manifestly heterogeneous.

21. The reason of this difficulty is, that our fundamental conceptions of lines and other relations of space, are capable of being contemplated under several various aspects, and more than one of these aspects are needed in our reasonings. We may take one such aspect of the conception for a definition; and then we must introduce the others by means of axioms. We may define parallels by their not meeting; but we must have some positive property, besides this negative one, in order to complete our reasonings respecting such lines. We have, in fact, our choice of several such self-evident properties, any of which we may employ for our purpose, as geometers well know; but with our naked definition, as they also know, we cannot proceed to the end. And in other cases, in like manner, our fundamental conception gives rise to various elementary truths, the connexion of which is the basis of our reasonings: but this connexion resides in our thoughts, and cannot be made to follow, as a logical result, from any assumed form of words, presented as a definition.

22. If it be further demanded, What is the nature of this bond in our thoughts by which various properties of lines are connected? perhaps the simplest answer is to say, that it resides in *the idea of space*. We cannot conceive things in space without being led to consider them as determined and related in some way or other to straight lines, right angles, and the like; and we cannot contemplate these determinations and relations distinctly, without assuming those properties of straight lines, of right angles, and of the rest, which are the basis of our Geometry. We cannot conceive or perceive objects at all, except as existing in space; we cannot contemplate them geometrically, without conceiving them in space which is subjected to

geometrical conditions; and this mode of contemplation is, by language, analysed into definitions, axioms, or both.

23. The truths thus seen and known, may be said to be known by *intuition*. In English writers this term has, of late, been vaguely used, to express all convictions which are arrived at without conscious reasoning, whether referring to relations among our perceptions, or to conceptions of the most derivative and complex nature. But if we were allowed to restrict the use of this term, we might conveniently confine it to those cases in which we necessarily apprehend relations of things truly, as soon as we conceive the objects distinctly. In this sense axioms may be said to be known *by intuition;* but this phraseology is not essential to our purpose.

24. It appears, then, that the evidence of the axioms of Geometry depends upon a distinct possession of the idea of space. These axioms are stated in the beginning of our Treatises, not as something which the reader is to learn, but as something which he already knows. No proof is offered of them; for they are the beginning not the end of demonstrations. The student's clear apprehension of the truth of these is a condition of the possibility of his pursuing the reasonings on which he is invited to enter. Without this mental capacity, and the power of referring to it, in the reader, the writer's assertions and arguments are empty and unmeaning words; but then, this capacity and power are what all rational creatures alike possess, though habit may have developed it in very various degrees in different persons.

25. It has been common in the school of metaphysicians of which I have spoken, to describe some of the elementary convictions of our minds as *fundamental laws* of belief; and it appears to have been considered that this might be taken as a final and sufficient account of such convictions. I do not know whether any persons would be tempted to apply this formula, as a solution of our question respecting the nature of axioms. If this were proposed, I should observe, that this form of expression seems to me, in such a case, highly unsatisfactory. For *laws* require and enjoin a conjunction of things which can be contemplated separately, and which would be disjoined if the law did not exist. It is a law of nature that terrestrial bodies, when free, fall downwards; for we can easily conceive such bodies divested of such a property. But we cannot say, in the same sense,

that the impossibility of two straight lines inclosing a space arises from a law; for if they are straight lines, they need no law to compel this result. We cannot conceive straight lines exempt from such a law. To speak of this property as imposed by a law, is to convey an inadequate and erroneous notion of the close necessity, inviolable even in thought, by which the truth clings to the conception of the lines.

26. This expression, of "laws of belief," appears to have found favour, on this account among others, that it recognised a kind of analogy between the grounds of our reasoning on very abstract subjects, and the principles to which we have recourse in other cases when we manifestly derive our fundamental truths from facts, and when it is supposed to be the ultimate and satisfactory account of them to say, that they are laws of nature learnt by observation. But such an analogy can hardly be considered as a real recommendation by the metaphysician; since it consists in taking a case in which our knowledge is obviously imperfect and its grounds obscure, and in erecting this case into an authority which shall direct the process and control the enquiry of a much more profound and penetrating kind of speculation. It cannot be doubted that we are likely to see the true grounds and evidence of our doctrines much more clearly in the case of Geometry and other rigorous systems of reasoning, than in collections of mere empirical knowledge, or of what is supposed to be such. It is both an unphilosophical and an indolent proceeding, to take the latter cases as a standard for the former.

27. I shall therefore consider it as established, that in Geometry our reasoning depends upon axioms as well as definitions,—that the evidence of the truth of the axioms and of the propriety of the definitions resides in the idea of space,—and that the distinct possession of this idea, and the consequent apprehension of the truth of the axioms which are its various aspects, is supposed in the student who is to pursue the path of geometrical reasoning. This being understood, I have little further to observe on the subject of Geometry. I will only remark—that all the conclusions which occur in the science follow purely from those first principles of which we have spoken;—that each proposition is rigorously proved from those which have been proved previously from such principles;—that this process of

successive proof is termed *Deduction;*—and that the rules which secure the rigorous conclusiveness of each step are the rules of *Logic,* which I need not here dwell upon.

28. But I now proceed to consider some other questions to which our examination of the evidence of Geometry was intended to be preparatory;—How far do the statements hitherto made apply to other sciences? for instance, to such sciences as are treated of in the present volume, Mechanics and Hydrostatics. To this I reply, that some such sciences at least, as for example Statics, appear to me to rest on foundations exactly similar to Geometry:—that is to say, that they depend upon axioms,—self-evident principles, not derived in any immediate manner from experiment, but involved in the very nature of the conceptions which we must possess, in order to reason upon such subjects at all. The proof of this doctrine must consist of several steps, which I shall take in order.

29. In the first place, I say that the axioms of Statics are *self-evidently true.* In the beginning of the preceding Treatise I have stated these barely as axioms, without addition or explanation, as the axioms of Geometry are stated in treatises on that subject. And such is the proper and orderly mode of exhibiting axioms; for, as has been said, they are to be understood as an expression of the condition of conception of the student. They are not to be learnt from without, but from within. They necessarily and immediately flow from the distinct possession of that idea, which if the student do not possess distinctly, all conclusive reasoning on the subject under notice is impossible. It is not the business of the deductive reasoner to communicate the apprehension of these truths, but to deduce others from them. . . .

. . . . . . . . . . . . .

40. Some persons may be disposed at first to say, that our knowledge of such elementary truths as are stated in the Axioms of Statics and Hydrostatics, is collected *from observation and experience.* But in refutation of this I remark, that we cannot experimentally verify these elementary truths, without assuming other principles which require proof as much as these do. If, for instance, Archimedes had wished to ascertain by trial whether two equal weights at the equal

arms of a lever would balance each other, how could he know that the weights *were* equal, by any more simple criterion than that they *did* balance? But in fact, it is perfectly certain that of the thousands of persons who from the time of Archimedes to the present day have studied Statics as a mathematical science, a very few have received or required any confirmation of his axioms from experiment; and those who have needed such help have undoubtedly been those in whom the apprehension of the real nature and force of the evidence of the subject was most obscure....

. . . . . . . . . . . . .

42. Thus, Statics, like Geometry, rests upon axioms which are neither derived directly from experience, nor capable of being superseded by definitions, nor by simpler principles. In this science, as in that previously considered, the evidence of these fundamental truths resides in those convictions, to which an attentive and steady consideration of the subject necessarily leads us. The axioms with regard to pressures, action and re-action, equilibrium and preponderance, rigid and flexible bodies, result necessarily from the conceptions which are involved in all exact reasoning on such matters. The axioms do not flow *from* the definitions, but they flow irresistibly *along with* the definitions, from the distinctness of our ideas upon the subjects thus brought into view. These axioms are not arbitrary assumptions, nor selected hypotheses; but truths which we must see to be necessarily and universally true, before we can reason on to anything else; and here, as in Geometry, the capacity of seeing that they are thus true, is required in the student, in order that he and the writer may be able to proceed together.

43. It was stated that the Axioms of Geometry are derived from the idea of space; in like manner the Axioms of Statics are derived from the *idea of statical force* or *pressure,* and the *idea of body* or *matter,* which, as we have said, is correlative with the idea of force. We must possess distinctly this idea of force acting upon body and body sustaining force;—of body resisting, and while it resists, transmitting the action of force;—of body, with this mechanical property, in the various forms of straight line, lever, plane, solid, flexible line, flexible surface, and fluid; and if we possess distinctly the ideas thus

pointed out, the truth of the Axioms of Statics and Hydrostatics will be seen as self-evident, and we shall be in a condition to go on with the reasonings of the preceding Treatise, seeing both the cogency of the proof, and its necessary and independent character.

44. As the Axioms which are the basis of the Statics of Solids depend upon the idea of body, considered as transmitting force, so the axioms of Hydrostatics depend on the idea of a fluid, considered as a body which transmits pressure in all directions; or, as we may express it more briefly, upon the *idea of fluid pressure.* It is not enough to conceive a fluid as a body, the parts of which are perfectly moveable; for, as I have elsewhere observed,[3] "this definition cannot be a sufficient basis for the doctrines of the pressure of fluids; for how can we evolve, out of the mere notion of mobility, which includes no conception of force, the independent conception of pressure." But the conception of fluid as transmitting pressure, supplies us with the requisite axioms. The First Axiom of our Hydrostatics—that if a fluid be contained in a tube of which the two ends are similar and equal planes acted on by equal pressures, it will be kept in equilibrium—follows from the principle of sufficient reason, for there is no reason why either pressure should preponderate. If, for example, the curvature of the tube, or any such cause, affected the pressure at either end, this condition would be a limitation of the property of transmitting pressure in all directions, and would imply imperfect fluidity; whereas the fluidity is supposed to be *perfect.* And for the like reasons, we might assume as an *Axiom* the Third *Proposition* of the Hydrostatics, that fluids transmit pressure *equally* in all directions, from one part of their boundary to the other; for if the pressure transmitted were different according to the direction, this difference might be referred to some cohesion or viscosity of the fluid; and the fluidity might be made more perfect, by conceiving the difference removed. Therefore the proposition would be necessarily and evidently true of a perfect fluid.

45. But instead of laying down this axiom, I have taken the axiom that any part of a fluid which is in equilibrium, may be supposed to become rigid. This axiom leads immediately to the proposition, and it is, besides, of great use in all parts of Hydrostatics. If we had to reason concerning flexible bodies, we might conveniently and prop-

erly assume a corresponding axiom for them;—namely, that, of a flexible body which is in equilibrium, any part may be supposed to become rigid. And we might give a reason for this, by saying that rigidity implies forces which resist a tendency to change of form, when any such tendency occurs; but in a body which is in equilibrium, there is no tendency to change of form, and therefore the resisting forces vanish. It is of no consequence what forces *would* act *if there were* a stress to bend the body: since there *is not* any such stress, the rigidity is not called into play, and therefore it makes no difference whether we suppose it to exist or not. . . .

. . . . . . . . . . . . .

51. In the case of Mechanics, as in the case of Geometry, the distinctness of the idea is necessary to a full apprehension of the truth of the axioms; and in the case of mechanical notions it is far more common than in Geometry, that the axioms are imperfectly comprehended, in consequence of the want of distinctness and exactness in men's ideas. Indeed this indistinctness of mechanical notions has not only prevailed in many individuals at all periods, but we can point out whole centuries, in which it has been, so far as we can trace, universal. And the consequence of this was, that the science of Statics, after being once established upon clear and sound principles, again fell into confusion, and was not understood as an exact science for two thousand years, from the time of Archimedes to that of Galileo and Stevinus.

52. In order to illustrate this indistinctness of mechanical ideas, I shall take from an ancient Greek writer an attempt to solve a mechanical problem; namely, the Problem of the Inclined Plane. The following is the mode in which Pappus professes[4] to answer this question:—"To find the force which will support a given weight *A* upon an inclined plane."

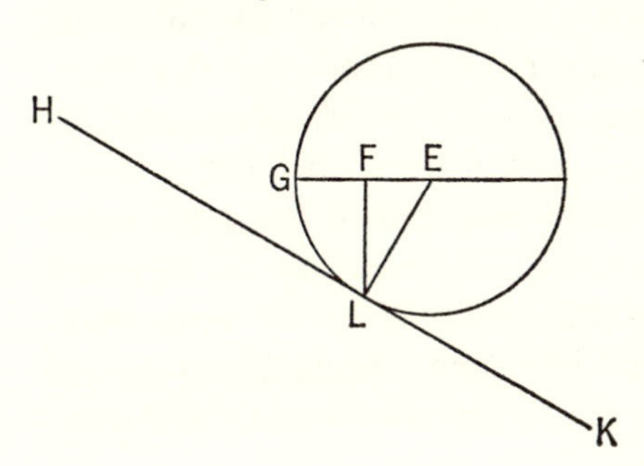


Let *HK* be the plane; let the weight *A* be formed into a sphere: let this sphere be placed in contact with the plane *HK*, touching it in the point *L*, and let *E* be its center. Let *EG* be a horizontal radius, and *LF* a vertical line which meets it. Take a weight *B* which is to *A* as *EF* to *FG*. Then if *A* and *B* be suspended at *E* and *G* to the lever *EFG* of which the center of motion is *F*, they will balance; being supported, as it were, by the fulcrum *LF*. And the sphere, which is equal to the weight *A*, may be supposed to be collected at its center. If therefore *B* act at *G*, the weight *A* will be supported.

It may be observed that in this attempt, the confusion of ideas is such, that the author assumes a weight which acts at *G*, on the lever *EFG*, and which is therefore a vertical force, as identical with a force which acts at *G*, to support the body in the inclined plane, and which is parallel to the plane.

53. When this kind of confusion was remedied, and when men again acquired distinct notions of pressure, and of the transmission of pressure from one point to another, the science of Statics was formed by Stevinus, Galileo, and their successors.[5]

The fundamental ideas of Mechanics being thus acquired, and the requisite consequences of them stated in axioms, our reasonings proceed by the same rigorous line of demonstration, and under the same logical rules as the reasonings of Geometry; and we have a science of Statics which is, like Geometry, an exact deductive science.

### *Sect. II. On the Logic of Induction*

54. There are other portions of Mechanics which require to be considered in another manner; for in these there occur principles which are derived directly and professedly from experiment and observation. The derivation of principles by reasoning from facts is performed by a process which is termed *Induction*, which is very different from the process of Deduction already noticed, and of which we shall attempt to point out the character and method.

It has been usual to say of any general truths, established by the consideration and comparison of several facts, that they are obtained by *Induction;* but the distinctive character of this process has not been well pointed out, nor have any rules been laid down which may prescribe the form and ensure the validity of the process, as has been done for Deductive reasoning by common Logic. The *Logic of Induction* has not yet been constructed; a few remarks on this subject are all that can be offered here.

55. The Inductive Propositions, to which we shall here principally refer as examples of their class, are those elementary principles which occur in considering the motion of bodies, and of which some are called the Laws of Motion. They are such as these;—a body not acted on by any force will move on for ever uniformly in a straight line;—gravity is a uniform force;—if a body in motion be acted upon by any force, the effect of the force will be compounded with the previous motion;—when a body communicates motion to another directly, the momentum lost by the first body is equal to the momentum gained by the second. And I remark, in the first place, that in collecting such propositions from facts, there occurs a step corresponding to the term "Induction," (ἐπαγωγὴ, *inductio*). Some notion is *superinduced* upon the observed facts. In each inductive process, there is some general idea introduced, which is given, not by the phenomena, but by the mind. The conclusion is not contained in the premises, but includes them by the introduction of a new generality. In order to obtain our inference, we travel beyond the cases we have before us; we consider them as exemplifications of, or deviations from, some ideal case in which the relations are complete and intelligible. We take a standard, and measure the facts by it; and this standard is created by us, not offered by Nature. Thus we assert, that a body left to itself will move on with unaltered velocity, not because our senses ever disclosed to us a body doing this, but because (taking this as our ideal case) we find that all actual cases are intelligible and explicable by means of the notion of forces which cause change of motion, and which are exerted by surrounding bodies. In like manner, we see bodies striking each other, and thus moving, accelerating, retarding, and stopping each other; but in all this, we do not, by our senses, perceive that abstract quantity, momentum, which is always lost by one as it is gained by another. This momentum is a creation of the mind, brought in among the facts, in order to convert their apparent confusion into order, their seeming chance into certainty, their perplexing variety into simplicity. This the idea of momentum gained and lost does; and, in like manner, in any other case in which inductive truths are established, some idea is introduced, as the means of passing from the facts to the truth.

56. The process of mind of which we here speak can only be described by suggestion and comparison. One of the most common of such comparisons, especially since the time of Bacon, is that which speaks of induction as the *interpretation* of facts. Such an expression is appropriate; and it may easily be seen that it includes the circumstance which we are now noticing;—the superinduction of an idea upon the facts by the interpreting mind. For when we read a page, we have before our eyes only black and white, form and colour; but by an act of the mind, we transform these perceptions into thought and emotion. The letters are nothing of themselves; they contain no truth, if the mind does not contribute its share: for instance, if we do not know the language in which the words are written. And if we are imperfectly acquainted with the language, we become very clearly aware how much a certain activity of the mind is requisite in order to convert the words into propositions, by the extreme effort which the business of interpretation requires. Induction, then, may be conveniently described as the interpretation of phenomena.

57. But I observe further, that in thus inferring truths from facts, it is not only necessary that the mind should contribute to the task its own idea, but, in order that the propositions thus obtained may have any exact import and scientific value, it is requisite that the idea be perfectly *distinct* and precise. If it be possible to obtain some vague apprehension of truths, while the ideas in which they are expressed remain indistinct and ill-defined, such knowledge cannot be available for the purposes we here contemplate. In order to construct a science, all our fundamental ideas must be distinct; and among them, those which Induction introduces.

58. This necessity for distinctness in the ideas which we employ in Induction, makes it proper to *define*, in precise and exact manner, each idea when it is thus brought forwards. Thus, in establishing the propositions which we have stated as our examples in these cases, we have to define *force* in general; *uniform force; compounding* of motions; *momentum*. The construction of these definitions is an essential part of the process of Induction, no less than the assertion of the inductive truth itself.

59. But in order to justify and establish the inference which we

make, the ideas which we introduce must not only be distinct, but also *appropriate.* They must be exactly and closely applicable to the facts; so that when the idea is in our possession, and the facts under our notice, we perceive that the former includes and takes up the latter. The idea is only a more precise mode of apprehending the facts, and it is empty and unmeaning if it be anything else; but if it be thus applicable, the proposition which is asserted by means of it is true, precisely because the facts *are* facts. When we have defined force to be the cause of change of motion, we see that, as we remove external forces, we do, in actual experiments, remove all the change of motion; and therefore the proposition that there is in bodies no internal cause of change of motion, is true. When we have defined momentum to be the product of the velocity and quantity of matter, we see that in the actions of bodies, the effect increases as the momentum increases; and by measurement, we find that the effect may consistently be measured by the momentum. The ideas here employed are not only distinct in the mind, but applicable in the world: they are the elements, not only of relations of thought, but of laws of nature.

60. Thus an inductive inference requires an idea from within, facts from without, and a coincidence of the two. The idea must be distinct, otherwise we obtain no scientific truth; it must be appropriate, otherwise the facts cannot be steadily contemplated by means of it; and when they are so contemplated, the Inductive Proposition must be seen to be verified by the evidence of sense.

It appears from what has been said, that in establishing a proposition by Induction, the definition of the idea and the assertion of the truth, are not only both requisite, but they are correlative. Each of the two steps contains the verification and justification of the other. The proposition derives its meaning from the definition; the definition derives its reality from the proposition. If they are separated, the definition is arbitrary or empty, the proposition is vague or verbal.

61. Hence we gather, that in the Inductive Sciences, our Definitions and our Elementary Inductive Truths ought to be introduced together. There is no value or meaning in definitions, except with reference to the truths which they are to express. Discussions about the definitions of any science, taken separately, cannot therefore be profitable,

if the discussion do not refer, tacitly or expressly, to the fundamental truths of the science; and in all such discussions it should be stated what are taken as the fundamental truths. With such a reference to Elementary Inductive Truths clearly understood, the discussion of Definitions may be the best method of arriving at that clearness of thought, and that arrangement of facts, which Induction requires.

I will now note some of the differences which exist between Inductive and Deductive Reasoning, in the modes in which they are presented.

62. One leading difference in these two kinds of reasoning is, that in Deduction we infer particular from general truths; in Induction, on the contrary, we infer general from particular. Deductive proofs consists of many steps, in each of which we apply known general propositions in particular cases;—"all triangles have their angles equal to two right angles, therefore this triangle has; therefore, &c." In Induction, on the other hand, we have a single step in which we pass from many particular Propositions to one general proposition; "This stone falls downwards; so do those others;—all stones fall downwards." And the former inference flows necessarily from the relation of general and particular; but the latter, as we have seen, derives its power of convincing from the introduction of a new idea, which is distinct and appropriate, and which supplies that generality which the particulars cannot themselves offer.

63. I observe also that this difference of process in inductive and deductive proofs, may be most properly marked by a difference in the form in which they are stated. In Deduction, the *Definition* stands at the beginning of the proposition; in Induction, it may most suitably stand at or near the end. Thus the definition of a uniform force is introduced in the course of the proposition that gravity is a uniform force. And this arrangement represents truly the real order of proof; for, historically speaking, it was taken for granted that gravity was a uniform force; but the question remained, what was the right definition of a uniform force. And in the establishment of other inductive principles, in like manner, definitions cannot be laid down for any useful purpose, till we know the propositions in which they are to be used. They may therefore properly come each at the conclusion of its corresponding proposition.

64. The ideas and definitions which are thus led to by our inductive process, may bring with them Axioms. Such Axioms may be self-evident as soon as the inductive idea has been distinctly apprehended, in the same manner as was explained respecting the fundamental ideas of Geometry and Statics. And thus *Axioms,* as well as Definitions, may come at the end of our Inductive Propositions; and they thus assume their proper place at the beginning of the deductive propositions which follow them, and are proved from them. . . .

. . . . . . . . . . . . .

65. Another peculiarity in inductive reasoning may be noticed. In a deductive demonstration, the reference is always to what has been already proved; in establishing an Inductive Principle, it is most convenient that the reference should be to subsequent propositions. For the proof of the Inductive Principle consists in this;—that the principle being adopted, consequences follow which agree with fact; but the demonstration of these consequences may require many steps, and several special propositions. Thus the Inductive Principle, that gravity is a uniform force, is established by shewing that the law of descent, which falling bodies follow in fact, is explained by means of this principle; namely, the law that the space is as the square of the time from the beginning of the motion. But the proof of such a property, from the definition of a uniform force, requires many steps . . . and this proof must be referred to, along with several others, in order to establish the truth, that gravity is a uniform force.

66. It may be suggested, that, this being the case, the propositions might be transposed, so that the inductive proof might come after those propositions to which it refers. But if this were done, all the propositions which depend upon the laws of motion must be proved hypothetically only. For instance, we must say, "If, in the communication of motion, the momentum lost and gained be equal, the velocity acquired by a body falling down an inclined plane, will be equal to that acquired by falling down the height." This would be inconvenient, and even if it were done, that completeness in the line of demonstration which is the object of the change, could not be obtained; for

the transition from the particular cases to the general truth, which must occur in the Inductive Proposition, could not be in any way justified according to rules of Deductive Logic.

I have, therefore, in the preceding pages, placed the Inductive Principle first in each line of reasoning; and have ranged after it the Deductions from it, which justify and establish it as their first office, but which are more important as its consequences and applications, after it is supposed to be established.

67. I have used one common *formula* in presenting the proof of each of the Inductive Principles which I have introduced;—namely, after stating or exemplifying the facts which the induction includes, I have added "These results can be clearly explained and rigorously deduced by introducing the *Idea* or the *Definition*," which belongs to each case, "and the *Principle*," which expresses the inductive truth. I do not mean to assert that this formula is the only right one, or even the best; but it appears to me to bring under notice the main circumstances which render an induction systematic and valid.

68. It may be observed, however, that this formula does not express the full cogency of the proof. It declares only that the results *can* be clearly explained and rigorously deduced by the employment of a certain definition and a certain proposition. But in order to make the conclusion demonstrative, we ought to be able to declare that the results can be clearly explained and rigorously deduced *only* by the definition and proposition which we adopt. And, in reality, the mathematician's conviction of the truth of the Laws of Motion does depend upon his seeing that they (or laws equivalent to them) afford the *only* means of clearly expressing and deducing the actual facts. But this conviction, that no other law than those proposed can account for the known facts, finds its place in the mind gradually, as the contemplation of the consequences of the law and the various relations of the facts becomes steady and familiar. I have therefore not thought it proper to require such a conviction along with the first assent to the inductive truths which I have here stated.

69. The propositions established by Induction are termed *Principles,* because they are the starting points of trains of deductive reasoning. In the system of deduction, they occupy the same place

as axioms; and accordingly they are termed so by Newton—"Axiomata sive leges motus." Stewart objects strongly to this expression: and it would be difficult to justify it; although to draw the line between axioms and inductive principles may be a harder task than at first appears.

70. But from the consideration that our Inductive Propositions are the principles or beginnings of our deductive reasoning, and so far at least stand in the place of axioms, we may gather *this* lesson,—that they are not to be multiplied without necessity. For instance, if in a treatise on Hydrostatics, we should state as two separate propositions, that "air has weight;" and that "the mercury in the barometer is sustained by the weight of the air;" and should prove both the one and the other by reference to experiment; we should offend against the maxims of Logic. These propositions are connected; the latter may be demonstrated deductively from the former; the former may be inferred inductively from the facts which prove the latter. One of these two courses ought to be adopted; we ought not to have two ends of our reasoning upwards, or two beginnings of our reasoning downwards.

71. I shall not now extend these Remarks further. They may appear to many barren and unprofitable speculations; but those who are familiar with such subjects, will perhaps find in them something which, if well founded, is not without some novelty for the English reader. Such will, I think, be the case, if I have satisfied him,—that mathematical truth depends on axioms as well as definitions,—that the evidence of geometrical axioms is to be found only in the distinct possession of the idea of space,—that other branches of mathematics also depend on axioms,—and that the evidence of these axioms is to be sought in some appropriate idea;—that the evidence of the axioms of statics, for instance, resides in the ideas of force and matter;—that in the process of induction the mind must supply an idea in addition to the facts apprehended by the senses;—that in each such process we must introduce one or more definitions, as well as a proposition;—that the definition and the proposition are correlative, neither being useful or valid without the other;—and that the formula of inductive reasoning must be in many respects the reverse of the common logical formulae of deduction.

# ON THE FUNDAMENTAL ANTITHESIS OF PHILOSOPHY*

1. All persons who have attended in any degree to the views generally current of the nature of reasoning are familiar with the distinction of *necessary* truths and *truths of experience;* and few such persons, or at least few students of mathematics, require to have this distinction explained or enforced. All geometricians are satisfied that the geometrical truths with which they are conversant are necessarily true: they not only are true, but they must be true. The meaning of the terms being understood, and the proof being gone through, the truth of the proposition must be assented to. That parallelograms upon the same base and between the same parallels are equal;—that angles in the same segment are equal;—these are propositions which we learn to be true by demonstrations deduced from definitions and axioms; and which, when we have thus learnt them, we see could not be otherwise. On the other hand, there are other truths which we learn from experience; as for instance, that the stars revolve round the pole in one day; and that the moon goes through her phases from full to full again in thirty days. These truths we see to be true; but we know them only by experience. Men never could have discovered them without looking at the stars and the moon; and having so learnt them, still no one will pretend to say that they are necessarily true. For aught we can see, things might have been otherwise; and if we had been placed in another part of the solar system, then, according to the opinions of astronomers, experience would have presented them otherwise.

2. I take the astronomical truths of experience to contrast with the geometrical necessary truths, as being both of a familiar definite

* From "On the Fundamental Antithesis of Philosophy," *Transactions of the Cambridge Philosophical Society,* VIII, pp. 170–81 (read 5 Feb., 1844).

sort; we may easily find other examples of both kinds of truth. The truths which regard numbers are necessary truths. It is a necessary truth, that 27 and 38 are equal to 65; that half the sum of two numbers added to half their difference is equal to the greater number. On the other hand, that sugar will dissolve in water; that plants cannot live without light; and in short, the whole body of our knowledge in chemistry, physiology, and the other inductive sciences, consists of truths of experience. If there be any science which offer to us truths of an ambiguous kind, with regard to which we may for a moment doubt whether they are necessary or experiential, we will defer the consideration of them till we have marked the distinction of the two kinds more clearly.

3. One mode in which we may express the difference of necessary truths and truths of experience, is, that necessary truths are those *of which we cannot distinctly conceive the contrary.* We can very readily conceive the contrary of experiential truths. We can conceive the stars moving about the pole or across the sky in any kind of curves with any velocities; we can conceive the moon always appearing during the whole month as a luminous disk, as she might do if her light were inherent and not borrowed. But we cannot conceive one of the parallelograms on the same base and between the same parallels larger than the other; for we find that, if we attempt to do this, when we separate the parallelograms into parts, we have to conceive one triangle larger than another, both having all their parts equal; which we cannot conceive at all, if we conceive the triangles distinctly. We make this impossibility more clear by conceiving the triangles to be placed so that two sides of the one coincide with two sides of the other; and it is then seen, that in order to conceive the triangles unequal, we must conceive the two bases which have the same extremities both ways, to be different lines, though both straight lines. This it is impossible to conceive: we assent to the impossibility as an axiom, when it is expressed by saying, that two straight lines cannot inclose a space; and thus we cannot distinctly conceive the contrary of the proposition just mentioned respecting parallelograms.

4. But it is necessary, in applying this distinction, to bear in mind the terms of it;—that we cannot *distinctly* conceive the contrary

of a necessary truth. For in a certain loose, indistinct way, persons conceive the contrary of necessary geometrical truths, when they erroneously conceive false propositions to be true. Thus, Hobbes erroneously held that he had discovered a means of geometrically doubling the cube, as it is called, that is, finding two mean proportionals between two given lines; a problem which cannot be solved by plane geometry. Hobbes not only proposed a construction for this purpose, but obstinately maintained that it was right, when it had been proved to be wrong. But then, the discussion showed how indistinct the geometrical conceptions of Hobbes were; for when his critics had proved that one of the lines in his diagram would not meet the other in the point which his reasoning supposed, but in another point near to it; he maintained, in reply, that one of these points was large enough to include the other, so that they might be considered as the same point. Such a mode of conceiving the opposite of a geometrical truth, forms no exception to the assertion, that this opposite cannot be distinctly conceived.

5. In like manner, the indistinct conceptions of children and of rude savages do not invalidate the distinction of necessary and experiential truths. Children and savages make mistakes even with regard to numbers; and might easily happen to assert that 27 and 38 are equal to 63 or 64. But such mistakes cannot make such arithmetical truths cease to be necessary truths. When any person conceives these numbers and their addition distinctly, by resolving them into parts, or in any other way, he sees that their sum is necessarily 65. If, on the ground of the possibility of children and savages conceiving something different, it be held that this is not a necessary truth, it must be held on the same ground, that it is not a necessary truth that 7 and 4 are equal to 11; for children and savages might be found so unfamiliar with numbers as not to reject the assertion that 7 and 4 are 10, or even that 4 and 3 are 6, or 8. But I suppose that no persons would on such grounds hold that these arithmetical truths are truths known only by experience.

6. Necessary truths are established, as has already been said, by demonstration, proceeding from definitions and axioms, according to exact and rigorous inferences of reason. Truths of experience are collected from what we see, also according to inferences of reason,

but proceeding in a less exact and rigorous mode of proof. The former depend upon the relations of the ideas which we have in our minds: the latter depend upon the appearances or phenomena, which present themselves to our senses. Necessary truths are formed from our thoughts, the elements of the world within us; experiential truths are collected from things, the elements of the world without us. The truths of experience, as they appear to us in the external world, we call Facts; and when we are able to find among our ideas a train which will conform themselves to the apparent facts, we call this a Theory.

7. This distinction and opposition, thus expressed in various forms; as Necessary and Experiential Truth, Ideas and Senses, Thoughts and Things, Theory and Fact, may be termed the *Fundamental Antithesis of Philosophy;* for almost all the discussions of philosophers have been employed in asserting or denying, explaining or obscuring this antithesis. It may be expressed in many other ways; but is not difficult, under all these different forms, to recognize the same opposition: and the same remarks apply to it under its various forms, with corresponding modifications. Thus, as we have already seen, the antithesis agrees with that of Reasoning and Observation: again, it is identical with the opposition of Reflection and Sensation: again, sensation deals with Objects; facts involve Objects, and generally all things without us are Objects:—Objects of sensation, of observation. On the other hand, we ourselves who thus observe objects, and in whom sensation is, may be called the Subjects of sensation and observation. And this distinction of Subject and Object is one of the most general ways of expressing the fundamental antithesis, although not yet perhaps quite familiar in English. I shall not scruple however to speak of the Subjective and Objective element of this antithesis, where the expressions are convenient.

8. All these forms of antithesis, and the familiar references to them which men make in all discussions, show the fundamental and necessary character of the antithesis. We can have no knowledge, except we have both impressions on our senses from the world without, and thoughts from our minds within:—except we attend to things, and to our ideas;—except we are passive to receive impressions, and active to compare, combine, and mould them. But on the

other hand, philosophy seeks to distinguish the impressions of our senses from the thoughts of our minds;—to point out the difference of ideas and things;—to separate the active from the passive faculties of our being. The two elements, sensations and ideas, are both requisite to the existence of our knowledge, as both matter and form are requisite to the existence of a body. But philosophy considers the matter and the form separately. The properties of the form are the subject of geometry, the properties of the matter are the subject of chemistry or mechanics.

9. But though philosophy considers these elements of knowledge separately, they cannot really be separated, any more than can matter and form. We cannot exhibit matter without form, or form without matter; and just as little can we exhibit sensations without ideas, or ideas without sensations;—the passive or the active faculties of the mind detached from each other.

In every act of my knowledge, there must be concerned the things whereof I know, and thoughts of me who know: I must both passively receive or have received impressions, and I must actively combine them and reason on them. No apprehension of things is purely ideal: no experience of external things is purely sensational. If they be conceived as *things,* the mind must have been awakened to the conviction of things by sensation: if they be *conceived* as things, the expressions of the senses must have been bound together by conceptions. If we *think* of any *thing,* we must recognize the existence both of thoughts and of things. *The fundamental antithesis of philosophy is an antithesis of inseparable elements.*

10. Not only cannot these elements be separately exhibited, but they cannot be separately conceived and described. The description of them must always imply their relation; and the names by which they are denoted will consequently always bear a relative significance. And thus *the terms which denote the fundamental antithesis of philosophy cannot be applied absolutely and exclusively in any case.* We may illustrate this by a consideration of some of the common modes of expressing the antithesis of which we speak. The terms Theory and Fact are often emphatically used as opposed to each other: and they are rightly so used. But yet it is impossible to say absolutely in any case, This is a Fact and not a Theory; this is a

Theory and not a Fact, meaning by Theory, true Theory. Is it a fact or a theory that the stars appear to revolve round the pole? Is it a fact or a theory that the sun attracts the earth? Is it a fact or a theory that a loadstone attracts a needle? In all these cases, some persons would answer one way and some persons another. A person who has never watched the stars, and has only seen them from time to time, considers their circular motion round the pole as a theory, just as he considers the motion of the sun in the ecliptic as a theory, or the apparent motion of the inferior planets round the sun in the zodiac. A person who has compared the measures of different parts of the earth, and who knows that these measures cannot be conceived distinctly without supposing the earth a globe, considers its globular form a fact, just as much as the square form of his chamber. A person to whom the grounds of believing the earth to revolve round its axis and round the sun, are as familiar as the grounds for believing the movements of the mail-coaches in this country, conceives the former events to be facts, just as steadily as the latter. And a person who, believing the fact of the earth's annual motion, refers it distinctly to its mechanical course, conceives the sun's attraction as a fact, just as he conceives as a fact the action of the wind which turns the sails of a mill. We see then, that in these cases we cannot apply absolutely and exclusively either of the terms, Fact or Theory. Theory and Fact are the elements which correspond to our Ideas and our Senses. The Facts are facts so far as the Ideas have been combined with the sensations and absorbed in them: the Theories are Theories so far as the Ideas are kept distinct from the sensations, and so far as it is considered as still a question whether they can be made to agree with them. A true Theory is a fact, a Fact is a familiar theory.

In like manner, if we take the terms Reasoning and Observation; at first sight they appear to be very distinct. Our observation of the world without us, our reasonings in our own minds, appear to be clearly separated and opposed. But yet we shall find that we cannot apply these terms absolutely and exclusively. I see a book lying a few feet from me: is this a matter of observation? At first, perhaps, we might be inclined to say that it clearly is so. But yet, all of us, who have paid any attention to the process of vision, and to the mode in which we are enabled to judge of the distance of objects, and to judge them to be distant objects at all, know that this judgment involves inferences drawn from various sensations;—from the impressions on our two eyes;—from our muscular sensations; and the like. These inferences are of the nature of reasoning, as much as when we judge of the distance of an object on the other side of a river by looking at it from different points, and stepping the distance between them. Or again: we observe the setting sun illuminate a gilded weathercock; but this is as much a matter of reasoning as when we observe the phases of the moon, and infer that she is illuminated by the sun. All observation involves inferences, and inference is reasoning.

11. Even the simplest terms by which the antithesis is expressed cannot be applied: ideas and sensations, thoughts and things, subject and object, cannot in any case be applied absolutely and exclusively. Our sensations require ideas to bind them together, namely, ideas of space, time, number, and the like. If not so bound together, sensations do not give us any apprehension of things or objects. All things, all objects, must exist in space and in time—must be one or many. Now space, time, number, are not sensations or things. They are something different from, and opposed to sensations and things. We have termed them ideas. It may be said they are *relations* of things, or of sensations. But granting this form of expression, still a *relation* is not a thing or a sensation; and therefore we must still have another and opposite element, along with our sensations. And yet, though we have thus these two elements in every act of perception, we cannot designate any portion of the act as absolutely and exclusively belonging to one of the elements. Perception involves sensation, along with ideas of time, space, and the like; or, if any one prefers the expression, involves sensations along with the apprehension of relations. Perception is sensation, along with such ideas as make sensation into apprehension of things or objects.

12. And as perception of objects implies ideas, as observation implies reasoning; so, on the other hand, ideas cannot exist where sensation has not been: reasoning cannot go on when there has not been previous observation. This is evident from the necessary order of development of the human faculties. Sensation necessarily exists

from the first moments of our existence, and is constantly at work. Observation begins before we can suppose the existence of any reasoning which is not involved in observation. Hence, at whatever period we consider our ideas, we must consider them as having been already engaged in connecting our sensations, and as modified by this employment. By being so employed, our ideas are unfolded and defined, and such development and definition cannot be separated from the ideas themselves. We cannot conceive space without boundaries or forms; now forms involve sensations. We cannot conceive time without events which mark the course of time; but events involve sensations. We cannot conceive number without conceiving things which are numbered; and things imply sensations. And the forms, things, events, which are thus implied in our ideas, having been the objects of sensation constantly in every part of our life, have modified, unfolded and fixed our ideas, to an extent which we cannot estimate, but which we must suppose to be essential to the processes which at present go on in our minds. We cannot say that objects create ideas; for to perceive objects we must already have ideas. But we may say, that objects and the constant perception of objects have so far modified our ideas, that we cannot, even in thought, separate our ideas from the perception of objects.

We cannot say of any ideas, as of the idea of space, or time, or number, that they are absolutely and exclusively ideas. We cannot conceive what space, or time, or number would be in our minds, if we had never perceived any thing or things in space or time. We cannot conceive ourselves in such a condition as never to have perceived any thing or things in space or time. But, on the other hand, just as little can we conceive ourselves becoming acquainted with space and time or numbers as objects of sensation. We cannot reason without having the operations of our minds affected by previous sensations; but we cannot conceive reasoning to be merely a series of sensations. In order to be used in reasoning, sensation must become observation; and, as we have seen, observation already involves reasoning. In order to be connected by our ideas, sensations must be things or objects, and things or objects already include ideas. And thus, as we have said, none of the terms by which the fundamental antithesis is expressed can be absolutely and exclusively applied.

13. I now proceed to make one or two remarks suggested by the views which have thus been presented. And first I remark, that since, as we have just seen, none of the terms which express the fundamental antithesis can be applied absolutely and exclusively, the absolute application of the antithesis in any particular case can never be a conclusive or immoveable principle. This remark is the more necessary to be borne in mind, as the terms of this antithesis are often used in a vehement and peremptory manner. Thus we are often told that such a thing is a *Fact* and not a Theory, with all the emphasis which, in speaking or writing, tone or italics or capitals can give. We see from what has been said, that when this is urged, before we can estimate the truth, or the value of the assertion, we must ask to whom is it a fact? what habits of thought, what previous information, what ideas does it imply, to conceive the fact as a fact? Does not the apprehension of the fact imply assumptions which may with equal justice be called theory, and which are perhaps false theory? in which case, the fact is no fact. Did not the ancients assert it as a fact, that the earth stood still, and the stars moved? and can any fact have stronger apparent evidence to justify persons in asserting it emphatically than this had? These remarks are by no means urged in order to show that no fact can be certainly known to be true; but only to show that no fact can be certainly shown to be a fact merely by calling it a fact, however emphatically. There is by no means any ground of general skepticism with regard to truth involved in the doctrine of the necessary combination of two elements in all our knowledge. On the contrary, ideas are requisite to the essence, and things to the reality of our knowledge in every case. The proportions of geometry and arithmetic are examples of knowledge respecting our ideas of space and number, with regard to which there is no room for doubt. The doctrines of astronomy are examples of truths not less certain respecting the external world.

14. I remark further, that since in every act of knowledge, observation or perception, both the elements of the fundamental antithesis are involved, and involved in a manner inseparable even in our conceptions, it must always be possible to derive one of these elements from the other, if we are satisfied to accept, as proof of such derivation, that one always co-exists with and implies the other.

Thus an opponent may say, that our ideas of space, time, and number, are derived from our sensations or perceptions, because we never were in a condition in which we had the ideas of space and time, and had not sensations or perceptions. But then, we may reply to this, that we no sooner perceive objects than we perceive them as existing in space and time, and therefore the ideas of space and time are not derived from the perceptions. In the same manner, an opponent may say, that all knowledge which is involved in our reasonings is the result of experience; for instance, our knowledge of geometry. For every geometrical principle is presented to us by experience as true; beginning with the simplest, from which all others are derived by processes of exact reasoning. But to this we reply, that experience cannot be the origin of such knowledge; for though experience shows that such principles are true, it cannot show that they *must be* true, which we also know. We never have seen, as a matter of observation, two straight lines inclosing a space; but we venture to say further, without the smallest hesitation, that we never shall see it; and if any one were to tell us that, according to his experience, such a form was often seen, we should only suppose that he did not know what he was talking of. No number of acts of experience can add to the certainty of our knowledge in this respect; which shows that our knowledge is not made up of acts of experience. We cannot test such knowledge by experience; for if we were to try to do so, we must first know that the lines with which we make the trial *are* straight; and we have no test of straightness better than this, that two such lines cannot inclose a space. Since then, experience can neither destroy, add to, nor test our axiomatic knowledge, such knowledge cannot be derived from experience. Since no one act of experience can affect our knowledge, no numbers of acts of experience can make it.

15. To this a reply has been offered, that it is a characteristic property of geometric forms that the ideas of them exactly resemble the sensations; so that these ideas are as fit subjects of experimentation as the realities themselves; and that by such experimentation we learn the truth of the axioms of geometry. I might very reasonably ask those who use this language to explain how a particular class of ideas can be said to resemble sensations; how, if they do, we can

know it to be so; how we can prove this resemblance to belong to geometrical ideas and sensations; and how it comes to be an especial characteristic of those. But I will put the argument in another way. Experiment can only show what is, not what must be. If experimentation on ideas shows what must be, it is different from what is commonly called experience.

I may add, that not only the mere use of our senses cannot show that the axioms of geometry *must be* true, but that, without the light of our ideas, it cannot even show that they *are* true. If we had a segment of a circle a mile long and an inch wide, we should have two lines inclosing a space; but we could not, by seeing or touching any part of either of them, discover that it was a bent line.

16. That mathematical truths are not derived from experience is perhaps still more evident, if greater evidence be possible, in the case of numbers. We assert that 7 and 8 are 15. We find it so, if we try with counters, or in any other way. But we do not, on that account, say that the knowledge is derived from experience. We refer to our conceptions of seven, of eight, and of addition, and as soon as we possess these conceptions distinctly, we see that the sum must be fifteen. We cannot be said to make a trial, for we should not believe the apparent result of the trial if it were different. If any one were to say that the multiplication table is a table of the results of experience, we should know that he could not be able to go along with us in our researches into the foundations of human knowledge; nor, indeed, to pursue with success any speculations on the subject.

17. Attempts have also been made to explain the origin of axiomatic truths by referring them to the association of ideas. But this is one of the cases in which the word *association* has been applied so widely and loosely, that no sense can be attached to it. Those who have written with any degree of distinctness on the subject, have truly taught, that the habitual association of the ideas leads us to believe a connexion of the things: but they have never told us that this association gave us the power of forming the ideas. Association may determine belief, but it cannot determine the possibility of our conceptions. The African king did not believe that water could become solid, because he had never seen it in that state. But that accident did not make it impossible to conceive it so, any more than

it is impossible for us to conceive frozen quicksilver, or melted diamond, or liquefied air; which we may never have seen, but have no difficulty in conceiving. If there were a tropical philosopher really incapable of conceiving water solidified, he must have been brought into that mental condition by abstruse speculations on the necessary relations of solidity and fluidity, not by the association of ideas.

18. To return to the results of the nature of the Fundamental Antithesis. As by assuming universal and indissoluble connexion of ideas with perceptions, of knowledge with experience, as an evidence of derivation, we may assert the former to be derived from the latter, so might we, on the same ground, assert the latter to be derived from the former. We see all forms in space; and we might hence assert all forms to be mere modifications of our idea of space. We see all events happen in time; and we might hence assert all events to be merely limitations and boundary-marks of our idea of time. We conceive all collections of things as two or three, or some other number; it might hence be asserted that we have an original idea of number, which is reflected in external things. In this case, as in the other, we are met at once by the impossibility of this being a complete account of our knowledge. Our ideas of space, of time, of number, however, distinctly reflected to us with limitations and modifications, must be reflected, limited and modified by something different from themselves. We must have visible or tangible forms to limit space, perceived events to mark time, distinguishable objects to exemplify number. But still, in forms, and events, and objects, we have a knowledge which they themselves cannot give us. For we know, without attending to them, that whatever they are, they will conform and must conform to the truths of geometry and arithmetic. There is an ideal portion in all our knowledge of the external world; and if we were resolved to reduce all our knowledge to one of its two antithetical elements, we might say that all our knowledge consists in the relation of our ideas. Wherever there is necessary truth, there must be something more than sensation can supply: and the necessary truths of geometry and arithmetic show us that our knowledge of objects in space and time depends upon necessary relations of ideas, whatever other element it may involve.

19. This remark may be carried much further than the domain of

geometry and arithmetic. Our knowledge of matter may at first sight appear to be altogether derived from the senses. Yet we cannot derive from the senses our knowledge of a truth which we accept as universally certain;—namely, that we cannot by any process add to or diminish the quantity of matter in the world. This truth neither is nor can be derived from experience; for the experiments which we make to verify it pre-suppose its truth. When the philosopher was asked what was the weight of smoke, he bade the inquirer subtract the weight of the ashes from the weight of the fuel. Every one who thinks clearly of the changes which take place in matter, assents to the justice of this reply: and this, not because any one had found by trial that such was the weight of the smoke produced in combustion, but because the weight lost was assumed to have gone into some other form of matter, not to have been destroyed. When men began to use the balance in chemical analysis, they did not prove by trial, but took for granted, as self-evident, that the weight of the whole must be found in the aggregate weight of the elements. Thus it is involved in the idea of matter that its amount continues unchanged in all changes which take place in its consistence. This is a necessary truth: and thus our knowledge of matter, as collected from chemical experiments, is also a modification of our idea of matter as the material of the world incapable of addition or diminution.

20. A similar remark may be made with regard to the mechanical properties of matter. Our knowledge of these is reduced, in our reasonings, to principles which we call the laws of motion. These laws of motion, as I have endeavoured to show,[1] depend upon the idea of Cause, and involve necessary truths, which are necessarily implied in the idea of cause;—namely, that every change of motion must have a cause—that the effect is measured by the cause;—that re-action is equal and opposite to action. These principles are not derived from experience. No one, I suppose, would derive from experience the principle, that every event must have a cause. Every attempt to see the traces of cause in the world assumes this principle. I do not say that these principles are anterior to experience; for I have already, I hope, shown, that neither of the two elements of our knowledge is, or can be, anterior to the other. But the two elements are co-ordinate in the development of the human mind; and the

ideal element may be said to be the origin of our knowledge with the more propriety of the two, inasmuch as our knowledge is the relation of ideas. The other element of knowledge, in which sensation is concerned, and which embodies, limits, and defines the necessary truths which express the relations of our ideas, may be properly termed experience; and I have, in the discussion just quoted, endeavoured to show how the principles concerning mechanical causation, which I have just stated, are, by observation and experiment, limited and defined, so that they become the laws of motion. And thus we see that such knowledge is derived from ideas, in a sense quite as general and rigorous, to say the least, as that in which it is derived from experience.

21. I will take another example of this; although it is one less familiar, and the consideration of it perhaps a little more difficult and obscure. The objects which we find in the world, for instance, minerals and plants, are of different kinds; and according to their kinds, they are called by various names, by means of which we know what we mean when we speak of them. The discrimination of these kinds of objects, according to their different forms and other properties, is the business of chemistry and botany. And this business of discrimination, and of consequent classification, has been carried on from the first periods of the development of the human mind, by an industrious and comprehensive series of observations and experiments; the only way in which any portion of the task could have been effected. But as the foundation of all this labour, and as a necessary assumption during every part of its progress, there has been in men's minds the principle, that objects are so distinguishable by resemblances and differences, that they may be named, and known by their names. This principle is involved in the idea of a Name; and without it no progress could have been made. The principle may be briefly stated thus:—Intelligible Names of kinds are possible. If we suppose this not to be so, language can no longer exist, nor could the business of human life go on. If instead of having certain definite kinds of minerals, gold, iron, copper and the like, of which the external forms and characters are constantly connected with the same properties and qualities, there were no connexion between the appearance and the properties of the object;—if what seemed externally

iron might turn out to resemble lead in its hardness; and what seemed to be gold during many trials, might at the next trial be found to be like copper; not only all the uses of these minerals would fail, but they would not be distinguishable kinds of things, and the names would be unmeaning. And if this entire uncertainty as to kind and properties prevailed for all objects, the world would no longer be a world to which language was applicable. To man, thus unable to distinguish objects into kinds, and call them by names, all knowledge would be impossible, and all definite apprehension of external objects would fade away into an inconceivable confusion. In the very apprehension of objects as intelligibly sorted, there is involved a principle which springs within us, contemporaneous, in its efficacy, with our first intelligent perception of the kinds of things of which the world consists. We assume, as a necessary basis of our knowledge, that things are of definite kinds; and the aim of chemistry, botany, and other sciences is, to find marks of these kinds; and along with these, to learn their definitely-distinguished properties. Even here, therefore, where so large a portion of our knowledge comes from experience and observation, we cannot proceed without a necessary truth derived from our ideas, as our fundamental principle of knowledge.

22. What the marks are, which distinguish the constant differences of kinds of things (definite marks, selected from among many unessential appearances), and what their definite properties are, when they are so distinguished, are parts of our knowledge to be learnt from observation, by various processes; for instance, among others, by chemical analysis. We find the differences of bodies, as shown by such analysis, to be of this nature:—that there are various elementary bodies, which, combining in different definite proportions, form kinds of bodies definitely different. But, in arriving at this conclusion, we introduce a new idea, that of Elementary Composition, which is not extracted from the phenomena, but supplied by the mind, and introduced in order to make the phenomena intelligible. That this notion of elementary composition is not supplied by the chemical phenomena of combustion, mixture, &c. as merely an observed fact, we see from this; that men had in ancient times performed many experiments in which elementary composition was concerned, and had not seen the fact. It never was truly seen till modern times; and when

seen, it gave a new aspect to the whole body of known facts. This idea of elementary composition, then, is supplied by the mind, in order to make the facts of chemical analysis and synthesis intelligible *as* analysis and synthesis. And this idea being so supplied, there enters into our knowledge along with it a corresponding necessary principle;—That the elementary composition of a body determines its kind and properties. This is, I say, a principle assumed, as a consequence of the idea of composition, not a result of experience; for when bodies have been divided into their kinds, we take for granted that the analysis of a single specimen may serve to determine the analysis of all bodies of the same kind: and without this assumption, chemical knowledge with regard to the kinds of bodies would not be possible. It has been said that we take only one experiment to determine the composition of any particular kind of body, because we have a thousand experiments to determine that bodies of the same kind have the same composition. But this is not so. Our belief in the principle that bodies of the same kind have the same composition is not established by experiments, but is assumed as a necessary consequence of the ideas of Kind and of Composition. If, in our experiments, we found that bodies supposed to be of the same kind had not the same composition, we should not at all doubt of the principle just stated, but conclude at once that the bodies were *not* of the same kind;—that the marks by which the kinds are distinguished had been wrongly stated. This is what has very frequently happened in the course of the investigations of chemists and mineralogists. And thus we have it, not as an experiential fact, but as a necessary principle of chemical philosophy, that the Elementary Composition of a body determines its Kind and Properties.

23. How bodies differ in their elementary composition, experiment must teach us, as we have already said, that experiment has taught us. But as we have also said, whatever be the nature of this difference, kinds must be definite, in order that language may be possible: and hence, whatever be the terms in which we are taught by experiment to express the elementary composition of bodies, the result must be conformable to this principle, That the differences of elementary composition are definite. The law to which we are led by experiment is, that the elements of bodies continue in definite propor-

tions according to weight. Experiments add other laws; as for instance, that of multiple proportions in different kinds of bodies composed of the same elements; but of these we do not here speak.

24. We are thus led to see that in our knowledge of mechanics, chemistry, and the like, there are involved certain necessary principles, derived from our ideas, and not from experience. But to this it may be objected, that the parts of our knowledge in which these principles are involved has, in historical fact, all been acquired by experience. The laws of motion, the doctrine of definite proportions, and the like, have all become known by experiment and observation; and so far from being seen as necessary truths, have been discovered by long-continued labours and trials, and through innumerable vicissitudes of confusion, error, and imperfect truth. This is perfectly true: but does not at all disprove what has been said. Perception of external objects and experience, experiment and observation are needed, not only, as we have said, to supply the objective element of all knowledge—to embody, limit, define, and modify our ideas; but this intercourse with objects is also requisite to unfold and fix our ideas themselves. As we have already said, ideas and facts can never be separated. Our ideas cannot be exercised and developed in any other form than in their combination with facts, and therefore the trials, corrections, controversies, by which the matter of our knowledge is collected, is also the only way in which the form of it can be rightly fashioned. Experience is requisite to the clearness and distinctness of our ideas, not because they are derived from experience, but because they can only be exercised upon experience. And this consideration sufficiently explains how it is that experiment and observation have been the means, and the only means, by which men have been led to a knowledge of the laws of nature. In reality, however, the necessary principles which flow from our ideas, and which are the basis of such knowledge, have not only been inevitably assumed in the course of such investigations, but have been often expressly promulgated in words by clear-minded philosophers, long before their true interpretation was assigned by experiment. This has happened with regard to such principles as those above mentioned; That every event must have a cause; That reaction is equal and opposite to action; That the quantity of matter in the world cannot be increased

or diminished: and there would be no difficulty in finding similar enunciations of the other principles above mentioned;—That the kinds of things have definite differences, and that these differences depend upon their elementary composition. In general, however, it may be allowed, that the necessary principles which are involved in those laws of nature of which we have a knowledge become then only clearly known, when the laws of nature are discovered which thus involve the necessary ideal element.

25. But since this is allowed, it may be further asked, how we are to distinguish between the necessary principle which is derived from our ideas, and the law of nature which is learnt by experience. And to this we reply, that the necessary principle may be known by the condition which we have already mentioned as belonging to such principles: . . . that it is impossible distinctly to conceive the contrary. We cannot conceive an event without a cause, except we abandon all distinct idea of cause; we cannot distinctly conceive two straight lines inclosing space; and if we seem to conceive this, it is only because we conceive indistinctly. We cannot conceive 5 and 3 making 7 or 9; if a person were to say that he could conceive this, we should know that he was a person of immature or rude or bewildered ideas, whose conceptions had no distinctness. And thus we may take it as the mark of a necessary truth, that we cannot conceive the contrary distinctly.

26. If it be asked what is the test of distinct conception (since it is upon the distinctness of conception that the matter depends), we may consider what answer we should give to this question if it were asked with regard to the truths of geometry. If we doubted whether any one had these distinct conceptions which enable him to see the necessary nature of geometrical truth, we should inquire if he could understand the axioms as axioms, and could follow, as demonstrative, the reasonings which are founded upon them. If this were so, we should be ready to pronounce that he had distinct ideas of space, in the sense now supposed. And the same answer may be given in any other case. That reasoner has distinct conceptions of mechanical causes who can see the axioms of mechanics as axioms, and can follow the demonstrations derived from them as demonstrations. If it be said that the science, as presented to him, may be erroneously constructed;

that the axioms may not be axioms, and therefore the demonstrations may be futile, we still reply, that the same might be said with regard to geometry: and yet that the possibility of this does not lead us to doubt either of the truth or of the necessary nature of the propositions contained in Euclid's Elements. We may add further, that although, no doubt, the authors of elementary books may be persons of confused minds, who present as axioms what are not axiomatic truths; yet that in general, what is presented as an axiom by a thoughtful man, though it may include some false interpretation or application of our ideas, will also generally include some principle which really is necessarily true, and which would still be involved in the axiom, if it were corrected so as to be true instead of false. And thus we still say, that if in any department of science a man can conceive distinctly at all, there are principles the contrary of which he cannot distinctly conceive, and which are therefore necessary truths.

27. But on this it may be asked, whether truth can thus depend upon the particular state of mind of the person who contemplates it; and whether that can be a necessary truth which is not so to all men. And to this we again reply, by referring to geometry and arithmetic. It is plain that truths may be necessary truths which are not so to all men, when we include men of confused and perplexed intellects; for to such men it is not a necessary truth that two straight lines cannot inclose a space, or that 14 and 17 are 31. It need not be wondered at, therefore, if to such men it does not appear a necessary truth that reaction is equal and opposite to action, or that the quantity of matter in the world cannot be increased or diminished. And this view of knowledge and truth does not make it depend upon the state of mind of the student, any more than geometrical knowledge and geometrical truth, by the confession of all, depend upon that state. We know that a man cannot have any knowledge of geometry without so much of attention to the matter of the science, and so much of care in the management of his own thoughts, as is requisite to keep his ideas distinct and clear. But we do not, on that account, think of maintaining that geometrical truth depends merely upon the state of the student's mind. We conceive that he knows it because it is true, not that it is true because he knows it. We are not surprised that attention and care and repeated thought should be requisite to the clear

apprehension of truth. For such care and such repetition are requisite to the distinctness and clearness of our ideas: and yet the relations of these ideas, and their consequences, are not produced by the efforts of attention or repetition which we exert. They are in themselves something which we may discover, but cannot make or change. The idea of space, for instance, which is the basis of geometry, cannot give rise to any doubtful propositions. What is inconsistent with the idea of space cannot be truly obtained from our ideas by any efforts of thought or curiosity; if we blunder into any conclusion inconsistent with the idea of space, our knowledge, as far as this goes, is no knowledge: any more than our observation of the external world would be knowledge, if, from haste or inattention, or imperfection of sense, we were to mistake the object which we see before us.

28. But further: not only has truth this reality, which makes it independent of our mistakes, that it must be what is really consistent with our ideas; but also, a further reality, to which the term is more obviously applicable, arising from the principle already explained, that ideas and perceptions are inseparable. For since, when we contemplate our ideas, they have been frequently embodied and exemplified in objects, and thus have been fixed and modified; and since this compound aspect is that under which we constantly have them before us, and free from which they cannot be exhibited; our attempts to make our ideas clear and distinct will constantly lead us to contemplate them as they are manifested in those external forms in which they are involved. Thus in studying geometrical truth, we shall be led to contemplate it as exhibited in visible and tangible figures;—not as if these could be sources of truth, but as enabling us more readily to compare the aspects which our ideas, applied to the world of objects, may assume. And thus we have an additional indication of the reality of geometrical truth, in the necessary possibility of its being capable of being exhibited in a visible or tangible form. And yet even this test by no means supersedes the necessity of distinct ideas, in order to a knowledge of geometrical truth. For in the case of the duplication of the cube by Hobbes, mentioned above, the diagram which he drew made two points appear to coincide, which did not really, and by the nature of our idea of space, coincide; and thus confirmed him in his error.

*Thus the inseparable nature of the Fundamental Antithesis of Ideas and Things gives reality to our knowledge, and makes objective reality a corrective of our subjective imperfections in the pursuit of knowledge. But this objective exhibition of knowledge can by no means supersede a complete development of the subjective condition, namely, distinctness of ideas. And that there is a subjective condition, by no means makes knowledge altogether subjective, and thus deprives it of reality; because, as we have said, the subjective and the objective elements are inseparably bound together in the fundamental antithesis.*

29. It would be easy to apply these remarks to other cases, for instance, to the case of the principle we have just mentioned, that the differences of elementary composition of different kinds of bodies must be definite. We have stated that this principle is necessarily true;—that the contrary proposition cannot be distinctly conceived. But by whom? Evidently, according to the preceding reasoning, by a person who distinctly conceives Kinds, as marked by intelligible names, and Composition, as determining the kinds of bodies. Persons new to chemical and classificatory science may not possess these ideas distinctly; or rather, cannot possess them distinctly; and therefore cannot apprehend the impossibility of conceiving the opposite of the above principle; just as the schoolboy cannot apprehend the impossibility of the numbers in his multiplication table being other than they are. But this inaptitude to conceive, in either case, does not alter the necessary character of the truth: although, in one case, the truth is obvious to all except schoolboys and the like, and the other is probably not clear to any except those who have attentively studied the philosophy of elementary compositions. At the same time, this difference of apprehension of the truth in different persons does not make the truth doubtful or dependent upon personal qualifications; for in proportion as persons attain to distinct ideas, they will see the truth; and cannot, with such ideas, see anything as truth which is not truth. When the relations of elements in a compound become as familiar to a person as the relations of factors in a multiplication table, he will then see what are the necessary axioms of chemistry, as he now sees the necessary axioms of arithmetic.

30. There is also one other remark which I will here make. In the progress of science, both the elements of our knowledge are

constantly expanded and augmented. By the exercise of observation and experiment, we have a perpetual accumulation of facts, the materials of knowledge, the objective element. By thought and discussion, we have a perpetual development of man's ideas going on: theories are framed, the materials of knowledge are shaped into form; the subjective element is evolved; and by the necessary coincidence of the objective and subjective elements, the matter and the form, the theory and the facts, each of these processes furthers and corrects the other: each element moulds and unfolds the other. Now it follows, from this constant development of the ideal portion of our knowledge, that we shall constantly be brought in view of new Necessary Principles, the expression of the conditions belonging to the Ideas which enter into our expanding knowledge. These principles, at first dimly seen and hesitatingly asserted, at last become clearly and plainly self-evident. Such is the case with the principles which are the basis of the laws of motion. Such may soon be the case with the principles which are the basis of the philosophy of chemistry. Such may hereafter be the case with the principles which are to be the basis of the philosophy of the connected and related polarities of chemistry, electricity, galvanism, magnetism. That knowledge is possible in these cases, we know; that our knowledge may be reduced to principles, gradually more simple, we also know; that we have reached the last stage of simplicity of our principles, few cultivators of the subject will be disposed to maintain; and that the additional steps which lead towards very simple and general principles will also lead to principles which recommend themselves by a kind of axiomatic character, those who judge from the analogy of the past history of science will hardly doubt. That the principles thus axiomatic in their form, do also express some relation of our ideas, of which experiment and observation have given a true and real interpretation, is the doctrine which I have here attempted to establish and illustrate in the most clear and undoubted of the existing sciences; and the evidence of this doctrine in those cases seems to be unexceptionable, and to leave no room to doubt that such is the universal type of the progress of science. Such a doctrine, as we have now seen, is closely connected with the views here presented of the nature of the Fundamental Antithesis of Philosophy, which I have endeavoured to illustrate.

Chapter ii

# The A Priori and the Empirical in Science

# ON THE NATURE OF THE TRUTH OF THE LAWS OF MOTION*

1. The long continuance of the disputes and oppositions of opinion which have occurred among theoretical writers concerning the elementary principles of Mechanics, may have made such discussions appear to some persons wearisome and unprofitable. I might, however, not unreasonably plead this very circumstance as an apology for offering a new view of the subject; since the extent to which these discussions have already gone shews that some men at least take a great interest in them; and it may be stated, I think, without fear of contradiction, that these controversies have not terminated in the general and undisputed establishment of any one of the antagonist opinions.

The question to which my remarks at present refer is this: "What is the kind and degree of cogency of the best proofs of the laws of motion, or of the fundamental principles of mechanics, exprest in any other way?" Are these laws, philosophically considered, *necessary*, and capable of demonstration by means of self-evident axioms, like the truths of geometry; or are they *empirical*, and only known to be true by trial and observation, like such general rules as we obtain in natural history?

It certainly appears, at first sight, very difficult to answer the arguments for either side of this alternative. On the one hand it is said, the laws of motion cannot be necessarily true, for if they were so, the denial of them would involve a contradiction. But this it does not, for we can readily conceive them to be other than they are. We can conceive that a body in motion should have a natural tendency to move slower and slower. And we know that, historically speaking,

* From "On the Nature of the Truth of the Laws of Motion," *Transactions of the Cambridge Philosophical Society*, V, pp. 149–72 (read 17 Feb., 1834).

men did at first suppose the laws of motion to be different from what they are now proved to be. This would have been impossible if the negation of these laws had involved a contradiction of self-evident principles, and consequently had been not only false but inconceivable. These laws, therefore, cannot be necessary; and can be duly established in no other way than by a reference to experience.

On the other hand, those who deduce their mechanical principles without any express reference to experiment, may urge, on their side, that, by the confession even of their adversaries, the laws of motion are proved to be true beyond the limits of experience;—that they are assumed to be true of any new kind of motion when first detected, as well as of those already examined;—and that it is inexplicable how such truths should be established empirically. They may add that the consequences of these laws are allowed to hold with the most complete and absolute universality; for instance, the proposition that "the quantity of motion in the world in a given direction cannot be either increased or diminished," is conceived to be rigorously exact; and to have a degree and kind of certainty beyond and above all mere facts of experience; what other kind of truth than necessary truth this can be, it is difficult to say. And if the conclusions be necessarily true, the principles must be so too.

This apparent contradiction therefore, that a law should be necessarily true and yet the contrary of it conceivable, is what I have now to endeavour to explain; and this I must do by pointing out what appear to me the true grounds of the laws of motion.

2. The science of Mechanics is concerned about motions as determined by their causes, namely, forces; the nature and extent of the truth of the first principles of this science must therefore depend upon the way in which we can and do reason concerning *causes*. In what manner we obtain the conception of cause, is a question for the metaphysician, and has been the subject of much discussion. But the general principle which governs our mode of viewing occurrences with reference to this conception, so far as our present subject is concerned, does not appear to be disturbed by any of the arguments which have been adduced in this controversy. This principle I shall state in the form of an axiom, as follows.

***Axiom I. Every change is produced by a cause***

It will probably be allowed that this axiom expresses a universal and constant conviction of the human mind; and that in looking at a series of occurrences, whether for theoretical or practical purposes, we inevitably and unconsciously assume the truth of this axiom. If a body at rest moves, or a body in motion stops, or turns to the right or the left, we cannot conceive otherwise than that there is some cause for this change. And so far as we can found our mechanical principles on this axiom, they will rest upon as broad and deep a basis as any truths which can come within the circle of our knowledge.

I shall not attempt to analyse this axiom further. Different persons may, according to their different views of such subjects, call it a law of our nature that we should think thus, or a part of the constitution of the human mind, or a result of our power of seeing the true relations of things. Such variety of opinion or expression would not affect the fundamental and universal character of the conviction which the axiom expresses; and would therefore not interfere with our future reasonings.

3. There is another axiom connected with this, which is also a governing and universal principle in all our reasoning concerning causes. It may be thus stated.

***Axiom II. Causes are measured by their effects***

Every effect, that is, every change in external objects, implies a cause, as we have already said: and the existence of the cause is known only by the effects it produces. Hence the intensity or magnitude of the cause cannot be known in any other manner than by these effects: and, therefore, when we have to assign a measure of the cause, we must take it from the effects produced.

In what manner the effects are to be taken into account, so as to measure the cause for any particular purpose, will have to be further considered; but the axiom, as now stated, is absolutely and universally true, and is acted upon in all parts of our knowledge in which causes are measured.

4. But something further is requisite. We not only consider that all changes of motion in a body have a cause, but that this cause may

reside in other bodies. Bodies are conceived to act upon one another, and thus to influence each other's motions, as when one billiard ball strikes another. But when this happens, it is also supposed that the body struck influences the motion of the striking body. This is included in our notion of body or matter. If one ball could strike and affect the motions of any number of others without having its own motion in any degree affected, the struck balls would be considered, not as bodies, but as mere shapes or appearances. Some reciprocal influence, some resistance, in short some *reaction*, is necessarily involved in our conception of action among bodies. All mechanical action upon matter implies a corresponding reaction; and we might describe matter as that which resists or reacts when acted on by force. Not only must there be a reaction in such cases, but this reaction is defined and determined by the action which produces it, and is of the same kind as the action itself. The action which one body exerts upon another is a blow, or a pressure; but it cannot press or strike without receiving a pressure or a blow in return. And the reciprocal pressure or blow depends upon the direct, and is determined altogether and solely by that. But this action being mutual, and of the same kind on each body, the effect on each body will be determined by the effect on the other, according to the same rule; each effect in turn being considered as action and the other as reaction. But this cannot be otherwise than by the equality and opposite direction of the action and reaction. And since this reasoning applies in all cases in which bodies influence each others' motions, we have the following axiom which is universally true, and is a fundamental principle with regard to all mechanical relations.

***Axiom III. Action is always accompanied by an equal and opposite Reaction***

5. I now proceed to shew in what manner the Laws of Motion depend upon these three axioms.

Bodies move in lines straight or curved, they move more or less rapidly, and their motions are variously affected by other bodies. This succession of occurrences suggests the conceptions of certain properties or attributes of the motions of bodies, as their direction and velocity, by means of which the laws of such occurrences may be exprest. And these properties or attributes are conceived as belonging to the body at each *point* of its motion, and as changing from one point to another. Thus the body, at each point of its path, moves in a certain direction, and with a certain velocity.

These properties, direction and velocity for instance, are subject to the rule stated in the first axiom: they cannot change without some cause; and when any changes in the motions of a body are seen to depend on its position relative to another body or to any part of space, such other body, or such other part of space, is said to exert a *force* upon the moving body. Also the force exerted upon the moving body is considered to be of a certain value at each point of the body's motion; and though it may change from one point to another, its changes must depend upon the position of the points only, and not upon the velocity and direction of the moving body. For the force which acts upon the body is conceived as a property of the bodies, or points, or lines, or surfaces among which the moving body is placed; the force at all points therefore depends upon the position with regard to the bodies and spaces of which the force is a property; but remains the same, whatever be the circumstances of the body moved. The circumstances of the body moved cannot be a cause which shall change the force acting at any point of space, although they may alter the *effect* which that force produces upon the body. Thus, gravity is the same force at the same point of space, whether it have to act upon a body at rest or in motion; although it still remains to be seen whether it will produce the same effect in the two cases.

6. This being established, we can now see of what nature the laws of motion must be, and can state in a few words the proofs of them. We shall have a law of motion corresponding to each of the above three axioms; the first law will assert that when no force acts, the properties of the motion will be constant; the second law will assert that when a force acts, its quantity is measured by the effect produced; the third law will assert that, when one body acts upon another, there will be a reaction, equal and opposite to the action. And so far as the laws are announced in this form, they will be of absolute and universal truth, and independent of any particular experiment or observation whatever.

But though these laws of motion are necessarily and infallibly true, they are, in the form in which we have stated them, entirely useless and inapplicable. It is impossible to deduce from them any definite and positive conclusions, without some additional knowledge or assumption. This will be clear by stating, as we can now do in a very small compass, the proofs of the laws of motion in the form in which they are employed in mechanical reasonings.

7. First, of the first Law;—that *a body not acted upon by any force will go on in a straight line with an invariable velocity.*

The body will go on in a straight line: for, at any point of its motion, it has a certain direction, which direction will, by Axiom I, continue unchanged, except some cause make it deviate to one side or other of its former position. But any cause which should make the direction deviate towards any part of space would be a force, and the body is not acted upon by any force. Therefore, the direction cannot change, and the body will go on in the same straight line from the first.

The body will move with an invariable velocity. For the velocity at any point will, by Axiom I, continue unchanged, except some cause make it increase or decrease. And since, by supposition, the body is not acted upon by any force, there can be no such cause depending upon position, that is, upon relations of *space;* for any cause of change of motion which has a reference to space is force.

Therefore there can be no cause of change of motion, except there be one depending upon *time,* such, for instance, as would exist if bodies had a natural tendency to move slower and slower, according to a rate depending on the time elapsed.

But if such cause existed, its effects ought to be considered separately; and it would still be requisite to assume the permanence of the same velocity, as the first law of motion; and to obtain, in addition to this, the laws of the retardation depending on the time.

Whether there is any such cause of retardation in the actual motions of bodies, can be known only by a reference to experience; and by such reference it appears that there is no such cause of the diminution of velocity depending on time alone; and therefore that the first law of motion may, in all cases in which bodies are exempt from the action of external forces, be applied without any addition or correction depending upon the time elapsed.

It is not here necessary to explain at any length in what manner we obtain from experience the knowledge of the truth just stated, that there is not in the mere lapse of time any cause of the retardation of moving bodies. The proposition is established by shewing that in all the cases in which such a cause appears to exist, the cause of retardation resides in surrounding bodies and not in time alone, and is therefore an external force. And as this can be shown in every instance, there remains only the negation of all ground for the assumption of such a cause of retardation. We therefore reject it altogether.

Thus it appears that in proving the first law of motion, we obtain from our conception of cause the conviction that velocity will be uniform except some cause produce a change in it; but that we are compelled to have recourse to experience in order to learn that time alone is not a cause of change of velocity.

8. I now proceed to the second Law:—that *when a force acts upon a body in motion, the effect is the same as that which the same force produces upon a body at rest.*

This law requires some explanation. How is the effect produced upon a moving body to be measured, so that we may compare it with the effect upon a body at rest? The answer to this is, that we here take for the measure of the effect of the force, that motion which must be *compounded* with the motion existing before the change, in order to produce the motion which exists after the change: the rules for the composition of motion being established on independent grounds by the aid of definition alone. Thus if gravity act upon a body which is falling vertically, the effect of gravity upon the body is measured by the velocity *added* to that which the body already has: if gravity act upon a body which is moving horizontally, its effect is measured by the distance to which the body falls below the horizontal line.

The effect of the force which we consider in the second Law of motion, is its effect upon velocity only: and it is proper to mark this restriction by an appropriate term: we shall call this the *accelerative effect* of force; and the cause, as measured by this effect, may be termed the *accelerative quantity* of the force.[1]

A law of motion which necessarily results from our second Axiom is, that the accelerative quantity of a force is measured by the accelerative effect. But whether the accelerative effect depends upon the velocity and direction of the moving body, cannot be known independently of experience. It is very conceivable, for instance, that the force of gravity being everywhere the same, shall yet produce, upon falling bodies, a smaller accelerative effect in proportion to the velocity which they already have in a downward direction. Indeed if gravity resembled in its operation the effect of any other mode of mechanical agency, the result would be so. If a body moved downwards in consequence of the action of a hand pushing it with a constant effort, or of a spring, or of a stream of fluid rushing in the same direction, the accelerative effect of such agents would be smaller and smaller as the velocity of the body propelled was larger and larger. We can learn from experience alone that the effects of the action of gravity do not follow the same rule.

We assert that the accelerative quantity of the *same force* of gravity is the same whatever be the motion of the body acted on. It may be asked how we know that the force of gravity *is* the same in cases so compared; for instance, when it acts on a body at rest and in motion? The answer to this question we have given already. By the very process of considering gravity as a force, we consider it as an attribute of something independent of the body acted on. The amount of the force may depend upon place, and even time, for any thing we know *a priori;* but we do not find that the weight of bodies depends on these circumstances, and therefore, having no evidence of a difference in the force of gravity, we suppose it the same at different times and places. And as to the rest, since the force is a force which acts on the body, it is considered as the same force, whatever be the circumstances of the passive body, although the *effects* may vary with these circumstances. If the effects are liable to such change, this change must be considered separately, and its laws investigated; but it cannot be allowed to unsettle our assumption of the permanence of the force itself. It is precisely this assumption of a constant cause, which gives us a fixed term, as a means of estimating and expressing by what conditions the effects are regulated.

It appears by observation and experiment, that the accelerative quantity of the same force is not affected by the velocity or direction of the body acted on: for instance, a body falling vertically receives, in any second of time, an accession of velocity as great as that which it received in the first second, notwithstanding the velocity with which it is already moving. The proof of this and similar assertions from experiment produced, historically speaking, the establishment of the second law of motion in the sense in which we now assert it. And here, as in the case of the first law, we may observe that an important portion of the process of proof consisted in shewing that in those cases in which the *accelerative effect* of a force appeared to be changed by the circumstances of the motion of the body acted on, the change was, in fact, due to other external forces; so that all evidence of a cause of change residing in those circumstances was entirely negatived; and thus the law, that the accelerative effect of the same force is the same, appeared to be absolutely and rigorously true.

9. When the motions of bodies are not effected merely by forces like gravity, which are only perceived by their effects, but are acted upon by other bodies, the case requires other considerations.

It is in such cases that we originally form the conception of force; we ourselves pull and push, thrust and throw bodies, with a view, it may be, either to put them in motion, or to prevent their moving, or to alter their figure. Such operations, and the terms by which they are described, are all included in the term *force,* and in other terms of cognate import. And in using this term, we necessarily assume and imply the co-existence of these various effects of force which we have observed universally to accompany each other. Thus the same kind of force which is the cause of motion, may also be the cause of a body having a form different from its natural form; when we draw a bow, the same kind of pull is needed to move the string, and to hold it steady, when the bow is bent. And a weight might be hung to the string, so as to produce either the one or the other of these effects. By an infinite multiplicity of experiments of this kind, we become imbued with the conviction that the same pressure may be the cause of tension and of motion. Also as the cause can be known by its effects only, each of these effects may be taken as its measure; and therefore, so long as one of them is the same, since the cause is the same, the other must be the same also. That is, so long as the pressure of force which shews itself in tension is the same, the motion which it would produce must, under the same circumstances, be the same also. This general fact is not a result of any particular observations, but of the general observation or suggestion arising unavoidably from universal experience, that both tension and motion may be referred to force as their cause, and have no other cause.

We come therefore to this principle with regard to the actions of bodies upon each other, that so long as the tension or pressure is the same, the force, as shewn by its effect in producing motion, must also be the same.

10. This force or action of bodies upon one another, is that which is meant in the Third Axiom, and we now proceed to consider the application of this axiom in mechanics.

Pressures or forces such as I have spoken of, may be employed in producing tension only, and not motion; in this case, each force prevents the motion which would be produced by the others, and the forces are said to balance each other, or to be in equilibrium. The science which treats of such cases is called Statics, and it depends entirely upon the above third axiom, applied to pressures producing rest. It follows from that axiom, that pressures, which acting in opposite directions thus destroy each other's effects, must be equal, each measuring the other. Thus if a man supports a stone in his hand, the force or effort exerted by the man upwards is equal to the weight or force of the stone downwards. And if a second stone, just equal to the first, were supported at the same time in the same hand, the force or effort must be twice as great; for the two stones may be considered as one body of twice the magnitude, and of twice the weight; and therefore the effort which supports it must also be twice as great. And thus we see in what manner statical forces are to be measured in virtue of this third axiom; and no further principle is requisite to enable us to establish the whole doctrine of statics.

11. The third axiom, when applied to the actions of bodies in motion, gives rise to the third law of motion, which we must now consider. Here, as in the cases of the other axioms, we must inquire how we are to measure the quantities to which the axiom applies. What is the measure of the *action* which takes place when a body is put in motion by pressure or force? In order to answer this question, we must consider what circumstances make it requisite that the

force should be greater or less. If we have to lift a stone, the force which we exert must be greater when the stone is greater: again, we must exert a greater force to lift it quickly than slowly. It is clear, therefore, that that property of a force with which we are here concerned, and which we may call the *motive quantity* of the force,[2] increases both when the velocity communicated, and when the mass moved, increase, and depends upon both these quantities, though we have not yet shewn what is the law of this dependence.

The condition that a quantity $P$ shall increase when each of two others $V$ and $M$ does so, may be satisfied in many ways: for instance, by supposing $P$ proportional to the sum $M+V$, (all the quantities being expressed in numbers), or to the product, $MV$, or to $MV^2$, or in many other ways.

When, however, the quantities $V$ and $M$ are altogether heterogeneous, as when one is velocity, and the other weight, the first of the above suppositions, that $P$ varies as $M+V$, is inadmissible. For the law of variation of the formula $M+V$ depends upon the relation of the units by which $M$ and $V$ respectively are measured; and as these units are arbitrary in each case, the result is, in like manner, arbitrary, and therefore cannot express a law of nature.

12. The supposition that the motive quantity of a force varies as $M + V$, where $M$ is the mass moved and $V$ the velocity, being thus inadmissible, we have to select upon due grounds, among the other formulae $MV$, $MV^2$, $M^2V$, &c.

And in the first place I observe that the formula must be proportional to $M$ simply (excluding $M^2$, &c.) for both the forces which produce motion and the masses in which motion is produced are capable of addition by juxtaposition, and it is easily seen by observation that such addition does not modify the motion of each mass. If a certain pressure upon one brick (as its own weight) cause it to fall with a certain velocity, an equal pressure on another equal brick will cause it also to fall with the same velocity; and these two bricks being placed in contact, may be considered as one mass, which a double force will cause to fall with still the same velocity. And thus all bodies whatever be their magnitude, will fall with the same velocity by the action of gravity. Those who deny this (as the Aristotelians did) must maintain, that by establishing between two bodies such a contact as makes them one body, we modify the motion which a certain pressure will produce in them. And when we find experimentally (as we do find) that large bodies and small ones fall with the same velocity, excluding the effects of extraneous forces, this result shews that there is not, in the union of small bodies into a larger one, any cause which affects the motion produced in the bodies.

It appears, therefore, that the motive quantity of force which puts a body in motion is, *cæteris paribus,* proportional to the mass of the body; so that for a double mass a double force is requisite, in order that the velocity produced may be the same. Mass considered with reference to this rule, is called *Inertia.*

13. The measure of mass which is used in expressing a law of motion, must be obtained in some way independent of motion, otherwise the law will have no meaning. Therefore, mass measured in order to be considered as *Inertia* must be measured by the statical effects of bodies, for instance, by comparison of weights. Thus two masses are equal which each balance the same weight in the same manner; and a mass is double of one of them which produces the same effect as the two. And we find, by universal observations, that the weight of a mass is not affected by the figure or the arrangement of parts, so long as the latter continues the same. Hence it appears that the mass of bodies must be compared by comparing their weights, and Inertia is proportional to weight at the same place.

Since all bodies, small or large, light or heavy, fall downwards with equal velocities, when we remove or abstract the effect of extraneous circumstances, the motive quantity of the force of gravity on equal bodies is as their masses; or as their weight, by what has just been said.

14. For the measure of the motive quantity of force, or of the action and reaction of bodies in motion, we have, therefore, now to choose among such expressions as $MV$, and $MV^2$. And our choice must be regulated by finding what is the measure which will enable us to assert, in all cases of action between bodies in motion, that action and reaction are equal and opposite.

Now the fact is, that either of the above measures may be taken, and each has been taken by a large body of mathematicians. The former however ($MV$) has obtained the designation which naturally falls to the lot of such a measure; and is called *momentum*, or sometimes simply *quantity of motion:* the latter quantity ($MV^2$) is called *vis viva* or *living force.*

I have said that either of these measures may be taken: the former must be the measure of action, if we are to measure it by the effect produced *in a given time;* the latter is the measure if we take the *whole* effect produced. In either way the third law of motion would be true.

Thus if a ball *B*, lying on a smooth table, be drawn along by a weight *A* hanging by a thread over the edge of the table, the motion of *B* is produced by the action of *A*, and on the other hand the motion of *A* is diminished by the reaction of *B*; and the equality of action and reaction here consists in this, that the momentum ($MV$) which *B* acquires in any time is equal to that which *A* loses: that is, so much is taken from the momentum which *A* would have had, if it had fallen freely *in the same time;* so that *A* falls more slowly by just so much.

But if the weight *A* fall through a given space from rest, as 1 foot, and then cease to act, the equality of action and reaction consists in this, that the *vis viva* which *B* acquires on the whole, is equal to the *vis viva* which *A* loses; that is, the *vis viva* of *A* thus acting on *B* is smaller by so much than it would have been, if *A* had fallen freely *through the same space.*

15. In fact, these two propositions are necessarily connected, and one of them may be deduced from the other. The former way of stating the third law of motion appears, however, to be the simplest mode of treating the subject, and we may put the third law of motion in this form.

*In the direct mutual action of bodies, the momentum gained and lost in any time are equal.*

This law depends upon experiment, and is perhaps best proved by some of its consequences. It follows from the law so stated, that the motive quantity of a force is proportional to the momentum generated in a given time; since the motive quantity of force is to be equivalent to that action and reaction which is understood in the third law of motion. Now, if the pressure arising from the weight of a body *P* produce motion in a mass *Q*, since the momentum gained by $Q$ and that lost by $P$ in any time are equal, the momentum of the whole at any time will be the same as if $P$'s weight had been employed in moving $P$ alone. Therefore, the velocity of the mass $Q$ will be less, in the same proportion in which the mass or inertia is greater: and thus the accelerating quantity of the force is inversely proportioned to the mass moved. This rule enables us to find the accelerative quantity of the force in various cases, as for instance, when bodies oscillate, or when a smaller weight moves a large mass; and we can hence calculate the circumstances of the motion, which are found to agree with the consequences of the above law.

16. But the argument may be reduced to a simpler form. Our object is to shew that, for an equal mass, the velocity produced by a force acting for a given time is as the pressure which produces the motion; for instance, that a double pressure will produce a double velocity. Now a double pressure may be considered as the union of two equal pressures, and if these two act *successively*, the first will communicate to the body a certain velocity, and the second will communicate an additional velocity, equal to the first, by the second law of motion; so that the whole velocity thus communicated will be the double of the first. Therefore, if the velocity communicated be not also the double of the first when the two pressures act *together*, the difference must arise from this, that the effect of one force is modified by the simultaneous action of the other. And when we find by experience (as we do find) that there is no such difference, but that the velocity communicated in a given time is as the pressure which communicates it, this result shews that there is nothing in the circumstance of a body being already acted on by one pressure, which modifies the effect of an additional pressure acting along with the first.

17. I have above asserted the law, of the *direct* action of bodies only. But it is also true when the action is indirect, as when by turning a winch we move a wheel, the main mass of which is farther from the axis than the handle of the winch. In this case the pressure we exert acts at a mechanical disadvantage on the main mass of the wheel, and we may ask whether this circumstance introduces any new law of motion. And to this we may reply, that we can *conceive* pressure to produce different effects in moving bodies, ac-

cording as it is exerted directly or by the intervention of machines; but that we *find* no reason to believe that such a difference exists. The relations of the pressures in different parts of a machine are determined by considering the machine at rest. But if we suppose it to be put in motion by such pressures, we see no reason to expect that these pressures should have a different relation to the motions produced from what they would have done if they were direct pressures. And as we find in experiment a negation of all evidence of such a difference, we reject the supposition altogether. We assert, therefore, the third law of motion to be true, whatever be the mechanism by the intervention of which action and reaction are opposed to each other.

From this consideration it is easy to deduce the following rule, which is known by the designation of D'Alembert's principle, and may be considered as a fourth law of motion.

*When any forces produce motion in any connected system of matter, the motive quantities of force gained and lost by the different parts must balance each other according to the connexion of the system.*

By the motive quantity of force *gained* by any body, is here meant the quantity by which that motive force which the body's motion implies (according to the measures already established) exceeds the quantity of motive force which acts immediately upon the body. It is the excess of the *effective* above the *impressed* force, and of course arises from the force transmitted from the other bodies of the system in consequence of the connexion of the parts. The motive quantity of force *lost* is in like manner the excess of the impressed above the effective force. And these two excesses, in different parts of the system, must balance each other according to the mechanical advantage or disadvantage at which they act for each part.

This completes our system of mechanical principles, and authorizes us to extend to bodies of any size and form the rules which the second law of motion gives for the motion of bodies considered as points. And by thus enabling us to trace what the motions of bodies will be according to the rule asserted in the third law of motion, (namely, that the motive quantity of forces is as the momentum produced in a given time,) it leads us to verify that supposition by experiments in which bodies oscillate or revolve or move in any regular and measurable manner, as has been done by Atwood, Smeaton, and many others.

18. We have thus a complete view of the nature and extent of the fundamental principles of mechanics; and we now see the reason why the laws of motion are so many and no more, in what way they are independent of experience, and in what way they depend upon experiment. The form, and even the language of these laws is of necessity what it is; but the interpretation and application of them is not possible without reference to fact. We may imagine many rules according to which bodies might move (for many sets of rules, different from the existing ones, are, so far as we can see, possible) and we should still have to assert—that velocity could not change without a cause,—that change of action is proportional to the force which produces it,—and that action and reaction are equal and opposite. The truth of these assertions is involved in those notions of causation and matter, which the very attempt to know any thing concerning the relations of matter and motion presupposes. But, according to the facts which we might find, in such imaginary cases as I have spoken of, we should settle in a different way—what is a cause of change of velocity,—what is the measure of the force which changes motion,—and what is the measure of action between bodies. The law is necessary, if there is to be a law; the meaning of its terms is decided by what we find, and is therefore regulated by our special experience.

19. It may further illustrate this matter to point out that this view is confirmed by the history of mathematics. The laws of motion were assented to as soon as propounded; but were yet each in its turn the subject of strenuous controversy. The terms of the law, the form, which is necessarily true, were recognized and undisputed; but the meaning of the terms, the substance of the law, was loudly contested; and though men often tried to decide the disputed points by pure reasoning, it was easily seen that this could not suffice; and that since it was a case where experience *could* decide, experience *must* be the proper test: since the matter came within her jurisdiction, her authority was single and supreme.

Thus with regard to the first law of motion, Aristotle allowed that *natural* motions continue unchanged, though he asserted the motions of terrestrial bodies to be *constrained* motions, and therefore, liable to diminution. Whether this was the cause of their diminution was a question of fact, which was, by examination of facts, decided against Aristotle. In like manner, in the first case of the second law of motion which came under consideration, both Galileo and his opponent agree that falling bodies are *uniformly* accelerated; that is, that the force of gravity accelerates a body uniformly whatever be the velocity it has already; but the question arises, what is uniform acceleration? It so happened in this case, that the first conjecture of Galileo, afterwards defended by Casræus, (that the velocity was proportional to the space from the beginning of the motion,) was not only contradictory to fact, but involved a self-contradiction; and was, therefore, easily disposed of. But this accident did not supersede the necessity of Galileo and his pupils verifying their assertion by reference to experiment, since there were many suppositions which were different from theirs, and still possible, though that of Casræus was not.

The mistake of Aristotle and his followers, in maintaining that large bodies fall more quickly than small ones, in exact proportion to their weight, arose from perceiving half of the third law of motion, that the velocity increases with the force which produces it; and from overlooking the remaining half, that a greater force is required for the same velocity, according as the mass is larger. The ancients never attained to any conception of the force which moves and the body which is moved, as distinct elements to be considered when we enquire into the subject of motion, and therefore could not even propose to themselves in a clear manner the questions which the third law of motion answered.

But, when, in more modern times, this distinction was brought into view, the progress of opinion in this case was nearly the same as with regard to the other laws.

It was allowed at once, and by all, that action and reaction are equal; but the controversy concerning the sense in which this law is to be interpreted, was one of the longest and fiercest in the history of mathematics, and the din of the war has hardly yet died away. The disputes concerning the measure of the force of bodies in motion, or the *vis viva,* were in fact a dispute which of two measures of action that I have mentioned above should be taken; the effect in a given time, or the whole effect: in the one case the *momentum* ($MV$), in the other the *vis viva* ($MV^2$), was the proper measure.

20. It may be observed that the word *momentum,* which one party appropriated to their views, was employed to designate the motive quantity of force, or the action of bodies in motion, before it was determined what the true measure of such action was. Thus Galileo, in his *Discorso intorno alle cose che stanno in su l'Acqua,* says, that momentum "is the force, efficacy, or virtue with which the motion moves and the body moved resists; depending not on weight only, but on the velocity, inclination, and any other cause of such virtue."

The adoption of the phrase *vis viva* is another instance of the extent to which men are tenacious of those terms which carry along with their use a reference to the fundamental laws of our thought on such matters. The party which used this phrase maintained that the mass multiplied into the square of the velocity was the proper measure of the force of bodies in motion; but finding the term *moving force* appropriated by their opponents, they still took the same term *force,* with the peculiar distinction of its being *living* force, in opposition to *dead* force or pressure, which they allowed to be rightly measured by the momentum generated in a given time. The same tendency to adopt, in a limited and technical sense, the words of most general and fundamental use in the subject, has led some writers (Newton for instance,) to employ the term *motion* or *quantity of motion* as synonymous with momentum, or the product of the numbers which express the mass and the velocity. And this use being established, the quantities of motion gained and lost are always equal and opposite; and, therefore the quantity which exists in any given direction cannot be increased or diminished by any mutual action of bodies. Thus we are led to the assertion which has already been noticed, that the quantity of motion in the world is always the same. And we now see how far the necessary truth of this proposition can be asserted. The proposition is necessarily true according to our notions of material causation; but the measure of "quantity of motion," which is a condition of its truth, is inevitably obtained from experience.

21. It is not surprizing that there should have been a good deal of confusion and difference of opinion on these matters: for it appears that there is, in the intellectual constitution and faculties of man, a source of self-delusion in such reasonings. The actual rules of the motion and mutual action of bodies are, and must be, obtained from observation of the external world: but there is a constant wish and propensity to express these rules in such terms as shall make them appear self-evident, because identical with the universal and necessary rules of causation. And this propensity is essential to the progress of our knowledge; and in the success of this effort consists, in a great measure, the advance of the science to its highest point of simplicity and generality.

22. The nature of the truth which belongs to the laws of motion will perhaps appear still more clearly, if we state, in the following tabular form, the analysis of each law into the part which is necessary, and the part which is empirical.

| | Necessary | Empirical |
|---|---|---|
| First Law | Velocity does not change without a cause | The time for which a body has already been in motion is not a cause of change of velocity. |
| Second Law | The accelerating quantity of a force is measured by the acceleration produced. | The velocity and direction of the motion which a body already possesses are not, either of them, causes which change the acceleration produced. |
| Third Law | Reaction is equal and opposite to action. | The connexion of the parts of a body, or of a system of bodies, and the action to which the body or system is already subject, are not, either of them, causes which change the effects of any additional action. |

Of course, it will be understood that, when we assert that the connexion of the parts of a system does not change the effect of any action upon it, we mean that this connexion does not introduce any *new* cause of change, but leaves the effect to be determined by the previously established rules of equilibrium and motion. The connexion will modify the application of such rules; but it introduces no additional rule: and the same observation applies to all the above stated empirical propositions.

This being understood, it will be observed that the part of each law which is here stated as empirical, consists, in each case, of a negation of the supposition that the condition of the moving body with respect to motion and action, is a cause of any change in the circumstances of its motion; and from this it follows that these circumstances are determined entirely by the forces extraneous to the body itself.

23. This mode of considering the question shows us in what manner the laws of motion may be said to be proved by their simplicity, which is sometimes urged as a proof. They undoubtedly have this distinction of the greatest possible simplicity, for they consist in the negation of all causes of change, except those which are essential to our conception of such causation. We may conceive the motions of bodies, and the effect of forces upon them, to be regulated by the lapse of time, by the motion which the bodies have, by the forces previously acting; but though we may imagine this as possible, we do not find that it is so in reality. If it were, we should have to consider the effect of these conditions of the body acted on, and to combine this effect with that of the acting forces; and thus the motion would be determined by more numerous conditions and more complex rules than those which are found to be the laws of nature. The laws which, in reality, govern motion are the fewest and simplest possible, because all are excluded, except those which the very nature of laws of motion necessarily implies. The prerogative of simplicity is possessed by the actual laws of the universe, in the highest perfection which is imaginable or possible. Instead of having to take into account all the circumstances of the moving bodies, we find that we have only to reject all these circumstances. Instead of having to combine empirical with necessary laws, we learn empirically that the necessary laws are entirely sufficient.

24. Since all that we learn from experience is, that she has nothing to teach us concerning the laws of motion, it is very natural that some persons should imagine that experience is not necessary to their proof. And accordingly many writers have undertaken to establish all the fundamental principles of machanics by reasoning alone. This has been done in two ways:—sometimes by attending only to the necessary part of each law (as the parts are stated in the last paragraph but one) and by overlooking the necessity of the empirical supplement and limitation to it;—at other times by asserting the part which I have stated as empirical to be self-evident, no less than the other part. The former way of proceeding may be found in many English writers on the subject; the latter appears to direct the reasonings of many eminent French mathematicians. Some (as Laplace) have allowed the empirical nature of two out of the three laws; others, as M. Poisson, have considered the first as alone empirical; and others; as D'Alembert, have assumed the self-evidence of all the three independently of any reference whatever to observation.

25. The parts of the laws which I have stated as empirical, appear to me to be clearly of a different nature, as to the cogency of their truth, from the parts which are necessary; and this difference is, I think, established by the fact that these propositions were denied, contested, and modified, before they were finally established. If these truths could not be denied without a self-contradiction, it is difficult to understand how they could be (as they were) long and obstinately controverted by mathematicians and others fully sensible to the cogency of necessary truth.

I will not however go so far as to assert that there may not be some point of view in which that which I have called the empirical part of these laws, (which, as we have seen, contains negatives only,) may be properly said to be self-evident. But however this may be, I think it can hardly be denied that there is a difference of a fundamental kind in the nature of these truths,—which we can, in our imagination at least, contradict and replace by others, and which, historically speaking, have been established by experiment;—and those other truths, which have been assented to from the first, and by all, and which we cannot deny without a contradiction in terms, or reject without putting an end to all use of our reason on this subject.

26. On the other hand, if any one should be disposed to maintain

that, inasmuch as the laws are interpreted by the aid of experience only, they must be considered as entirely empirical laws, I should not assert this to be placing the science of mechanics on a wrong basis. But at the same time I would observe, that the form of these laws is not empirical, and would be the same if the results of experience should differ from the actual results. The laws may be considered as a formula derived from *a priori* reasonings, where experience assigns the value of the terms which enter into the formula.

Finally, it may be observed, that if any one can convince himself that matter is either necessarily and by its own nature determined to move slower and slower, or necessarily and by its own nature determined to move uniformly, he must adopt the latter opinion, not only of the truth, but of the necessity of the truth of the first law of motion, since the former branch of the alternative is certainly false: and similar assertions may be made with regard to the other laws of motion.

27. This inquiry into the nature of the laws of motion, will, I hope, possess some interest for those who attach any importance to the logic and philosophy of science. The discussion may be said to be rather metaphysical than mechanical; but the views which I have endeavoured to present, appear to explain the occurrence and result of the principal controversies which the history of this science exhibits; and, if they are well founded, ought to govern the way in which the principles of the science are treated of, whether the treatise be intended for the mathematical student or the philosopher.

Chapter iii

# Induction and Scientific Method

# NOVUM ORGANON RENOVATUM*

# BOOK II

# OF THE CONSTRUCTION OF SCIENCE

## CHAPTER I
## OF TWO PRINCIPAL PROCESSES BY WHICH SCIENCE IS CONSTRUCTED

***Aphorism I***

*The two processes by which Science is constructed are the* Explication of Conceptions, *and the* Colligation of Facts.

To the subject of the present and next Book all that has preceded is subordinate and preparatory. In former works we have treated of the History of Scientific Discoveries and of the History of Scientific Ideas. We have now to attempt to describe the manner in which discoveries are made, and in which Ideas give rise to knowledge. It has already been stated that Knowledge requires us to possess both Facts and Ideas;—that every step in our knowledge consists in applying the Ideas and Conceptions furnished by our minds to the Facts which observation and experiment offer to us. When our Conceptions are clear and distinct, when our Facts are certain and sufficiently numerous, and when the Conceptions, being suited to the nature of the Facts, are applied to them so as to produce an exact and universal accordance, we attain knowledge of a precise and comprehensive kind, which we may term *Science*. And we apply this term to our knowledge still more decidedly when, Facts being

* From *Novum Organon Renovatum, being the Second part of the Philosophy of the Inductive Sciences,* 3rd Edition (London 1858).

thus included in exact and general Propositions, such Propositions are, in the same manner, included with equal rigour in Propositions of a higher degree of Generality; and these again in others of a still wider nature, so as to form a large and systematic whole.

But after thus stating, in a general way, the nature of science, and the elements of which it consists, we have been examining with a more close and extensive scrutiny, some of those elements; and we must now return to our main subject, and apply to it the results of our long investigation. We have been exploring the realm of Ideas; we have been passing in review the difficulties in which the workings of our own minds involve us when we would make our conceptions consistent with themselves: and we have endeavoured to get a sight of the true solutions of these difficulties. We have now to inquire how the results of these long and laborious efforts of thought find their due place in the formation of our Knowledge. What do we gain by these attempts to make our notions distinct and consistent; and in what manner is the gain of which we thus become possessed, carried to the general treasure-house of our permanent and indestructible knowledge? After all this battling in the world of ideas, all this struggling with the shadowy and changing forms of intellectual perplexity, how do we secure to ourselves the fruits of our warfare, and assure ourselves that we have really pushed forwards the frontier of the empire of Science? It is by such an appropriation, that the task which we have had in our hands during the two previous works, (the *History of the Inductive Sciences* and the *History of Scientific Ideas,*) must acquire its real value and true place in our design.

In order to do this, we must reconsider, in a more definite and precise shape, the doctrine which has already been laid down;—that our Knowledge consists in applying Ideas to Facts; and that the conditions of real knowledge are that the ideas be distinct and appropriate, and exactly applied to clear and certain facts. The steps by which our knowledge is advanced are those by which one or the other of these two processes is rendered more complete;—by which *Conceptions* are *made more clear* in themselves, or by which the Conceptions more strictly *bind together the Facts.* These two processes may be considered as together constituting the whole formation of our knowledge; and the principles which have been established in the History of Scientific Ideas bear principally upon the former of these two operations;—upon the business of elevating our conceptions to the highest possible point of precision and generality. But these two portions of the progress of knowledge are so clearly connected with each other, that we shall deal with them in immediate succession. And having now to consider these operations in a more exact and formal manner than it was before possible to do, we shall designate them by certain constant and technical phrases. We shall speak of the two processes by which we arrive at science, as *the Explication of Conceptions* and *the Colligation of Facts:* we shall show how the discussions in which we have been engaged have been necessary in order to promote the former of these offices; and we shall endeavour to point out modes, maxims, and principles by which the second of the two tasks may also be furthered.

## CHAPTER II
## OF THE EXPLICATION OF CONCEPTIONS

***Aphorism II***

*The Explication of Conceptions, as requisite for the progress of science, has been effected by means of discussions and controversies among scientists; often by debates concerning definitions; these controversies have frequently led to the establishment of a Definition; but along with the Definition, a corresponding Proposition has always been expressed or implied. The essential requisite for the advance of science is the clearness of the Conception, not the establishment of a Definition. The construction of an exact Definition is often very difficult. The requisite conditions of clear Conceptions may often be expressed by Axioms as well as by Definitions.*

***Aphorism III***

*Conceptions, for purposes of science, must be* appropriate *as well as clear: that is, they must be modifications of* that *Fundamental Idea, by which the phenomena can really be interpreted. This maxim may warn us from errour, though it may not lead to discovery.*

*Discovery depends upon the previous cultivation or natural clearness of the appropriate Idea, and therefore* no discovery is the work of accident.

### *Sect. I. Historical Progress of the Explication of Conceptions*

1. We have given the appellation of *Ideas* to certain comprehensive forms of thought,—as *space, number, cause, composition, resemblance,*—which we apply to the phenomena which we contemplate. But the special modifications of these ideas which are exemplified in particular facts, we have termed *Conceptions;* as *a circle, a square number, an accelerating force, a neutral combination* of elements, a *genus.* Such Conceptions involve in themselves certain necessary and universal relations derived from the Ideas just enumerated; and these relations are an indispensable portion of the texture of our knowledge. But to determine the contents and limits of this portion of our knowledge, requires an examination of the Ideas and Conceptions from which it proceeds. The Conceptions must be, as it were, carefully *unfolded,* so as to bring into clear view the elements of truth with which they are marked from their ideal origin. This is one of the processes by which our knowledge is extended and made more exact; and this I shall describe as the *Explication of Conceptions.*

In the several Books of the History of Ideas we have discussed a great many of the Fundamental Ideas of the most important existing sciences. We have, in those Books, abundant exemplifications of the process now under our consideration. We shall here add a few general remarks, suggested by the survey which we have thus made.

2. Such discussions as those in which we have been engaged concerning our fundamental Ideas, have been the course by which, historically speaking, those Conceptions which the existing sciences involve have been rendered so clear as to be fit elements of exact knowledge. Thus, the disputes concerning the various kinds and measures of *Force* were an important part of the progress of the science of Mechanics. The struggles by which philosophers attained a right general conception of *plane,* of *circular,* of *elliptical Polarization,* were some of the most difficult steps in the modern discoveries of Optics. A Conception of the *Atomic Constitution* of bodies, such as

shall include what we know, and assume nothing more, is even now a matter of conflict among Chemists. The debates by which, in recent times, the Conceptions of *Species* and *Genera* have been rendered more exact, have improved the science of Botany: the imperfection of the science of Mineralogy arises in a great measure from the circumstance, that in that subject, the Conception of a *Species* is not yet fixed. In Physiology, what a vast advance would that philosopher make, who should establish a precise, tenable, and consistent Conception of *Life!*

Thus discussions and speculations concerning the import of very abstract and general terms and notions, may be, and in reality have been, far from useless and barren. Such discussions arose from the desire of men to impress their opinions on others, but they had the effect of making the opinions much more clear and distinct. In trying to make others understand them, they learnt to understand themselves. Their speculations were begun in twilight, and ended in the full brilliance of day. It was not easily and at once, without expenditure of labour or time, that men arrived at those notions which now form the elements of our knowledge; on the contrary, we have, in the history of science, seen how hard, discoverers, and the forerunners of discoverers, have had to struggle with the indistinctness and obscurity of the intellect, before they could advance to the critical point at which truth became clearly visible. And so long as, in this advance, some speculators were more forward than others, there was a natural and inevitable ground of difference of opinion, of argumentation, of wrangling. But the tendency of all such controversy is to diffuse truth and to dispel errour. Truth is consistent, and can bear the tug of war; Errour is incoherent, and falls to pieces in the struggle. True Conceptions can endure the sun, and become clearer as a fuller light is obtained; confused and inconsistent notions vanish like visionary spectres at the break of a brighter day. And thus all the controversies concerning such Conceptions as science involves, have ever ended in the establishment of the side on which the truth was found.

3. Indeed, so complete has been the victory of truth in most of these instances, that at present we can hardly imagine the struggle to have been necessary. The very essence of these triumphs is that

they lead us to regard the views we reject as not only false, but inconceivable. And hence we are led rather to look back upon the vanquished with contempt than upon the victors with gratitude. We now despise those who, in the Copernican controversy, could not conceive the apparent motion of the sun on the heliocentric hypothesis;—or those who in opposition to Galileo, thought that a uniform force might be that which generated a velocity proportional to the space;—or those who held there was something absurd in Newton's doctrine of the different refrangibility of differently coloured rays;—or those who imagined that when elements combine, their sensible qualities must be manifest in the compound;—or those who were reluctant to give up the distinction of vegetables into herbs, shrubs, and trees. We cannot help thinking that men must have been singularly dull of comprehension, to find a difficulty in admitting what is to us so plain and simple. We have a latent persuasion that we in their place should have been wiser and more clear-sighted;—that we should have taken the right side, and given our assent at once to the truth.

4. Yet in reality, such a persuasion is a mere delusion. The persons who, in such instances as the above, were on the losing side, were very far, in most cases, from being persons more prejudiced, or stupid, or narrow-minded, than the greater part of mankind now are; and the cause for which they fought was far from being a manifestly bad one, till it had been so decided by the result of the war. It is the peculiar character of scientific contests, that what is only an epigram with regard to other warfare is a truth in this;—They who are defeated are really in the wrong. But they may, nevertheless, be men of great subtilty, sagacity, and genius; and we nourish a very foolish self-complacency when we suppose that we are their superiors. That this is so, is proved by recollecting that many of those who have made very great discoveries have laboured under the imperfection of thought which was the obstacle to the next step in knowledge. Though Kepler detected with great acuteness the Numerical Laws of the solar system, he laboured in vain to conceive the very simplest of the Laws of Motion by which the paths of the planets are governed. Though Priestley made some important steps in chemistry, he could not bring his mind to admit the doctrine of a

general Principle of Oxidation. How many ingenious men in the last century rejected the Newtonian Attraction as an impossible chimera! How many more, equally intelligent, have, in the same manner in our own time, rejected, I do not now mean as false, but as inconceivable, the doctrine of Luminiferous Undulations! To err in this way is the lot, not only of men in general, but of men of great endowments, and very sincere love of truth.

5. And those who liberate themselves from such perplexities, and who thus go on in advance of their age in such matters, owe their superiority in no small degree to such discussions and controversies as those to which we now refer. In such controversies, the Conceptions in question are turned in all directions, examined on all sides; the strength and the weakness of the maxims which men apply to them are fully tested; the light of the brightest minds is diffused to other minds. Inconsistency is unfolded into self-contradiction; axioms are built up into a system of necessary truths; and ready exemplifications are accumulated of that which is to be proved or disproved, concerning the ideas which are the basis of the controversy.

The History of Mechanics from the time of Kepler to that of Lagrange, is perhaps the best exemplification of the mode in which the progress of a science depends upon such disputes and speculations as give clearness and generality to its elementary conceptions. This, it is to be recollected, is the kind of progress of which we are now speaking; and this is the principal feature in the portion of scientific history which we have mentioned. For almost all that was to be done by reference to observation, was executed by Galileo and his disciples. What remained was the task of generalization and simplification. And this was promoted in no small degree by the various controversies which took place within that period concerning mechanical conceptions:—as, for example, the question concerning the measure of the Force of Percussion;—the war of the *Vis Viva;*—the controversy of the Center of Oscillation;—of the independence of Statics and Dynamics;—of the principle of Least Action;—of the evidence of the Laws of Motion;—and of the number of Laws really distinct. None of these discussions was without its influence in giving generality and clearness to the mechanical ideas of mathematicians: and therefore, though remote from general apprehension, and dealing with very abstract notions, they were of eminent use in the perfecting the science of Mechanics. Similar controversies concerning fundamental notions, those, for example, which Galileo himself had to maintain, were no less useful in the formation of the science of Hydrostatics. And the like struggles and conflicts, whether they take the form of controversies between several persons, or only operate in the efforts and fluctuations of the discoverer's mind, are always requisite, before the conceptions acquire that clearness which makes them fit to appear in the enunciation of scientific truth. This, then, was one object of the History of Ideas;—to bring under the reader's notice the main elements of the controversies which have thus had so important a share in the formation of the existing body of science, and the decisions on the controverted points to which the mature examination of the subject has led; and thus to give an abundant exhibition of that step which we term the Explication of Conceptions.

***Sect. II. Use of Definitions***

6. The result of such controversies as we have been speaking of, often appears to be summed up in a *Definition;* and the controversy itself has often assumed the form of a battle of definitions. For example, the inquiry concerning the Laws of Falling Bodies led to the question whether the proper Definition of a *uniform force* is, that it generates a velocity proportional to the *space* from rest, or to the *time*. The controversy of the *Vis Viva* was, what was the proper Definition of the *measure of force*. A principal question in the classification of minerals is, what is the Definition of a *mineral species*. Physiologists have endeavoured to throw light on their subject, by Defining *organization*, or some similar term.

7. It is very important for us to observe, that these controversies have never been questions of insulated and *arbitrary* Definitions, as men seem often tempted to suppose them to have been. In all cases there is a tacit assumption of some Proposition which is to be expressed by means of the Definition, and which gives it its importance. The dispute concerning the Definition thus acquires a real value, and becomes a question concerning true and false. Thus in the discussion of the question, What is a Uniform Force? it was taken for granted that 'gravity is a uniform force:'—in the debate of the

*Vis Viva*, it was assumed that 'in the mutual action of bodies the whole effect of the force is unchanged:'—in the zoological definition of Species, (that it consists of individuals which have, or may have, sprung from the same parents,) it is presumed that 'individuals so related resemble each other more than those which are excluded by such a definition;' or perhaps, that 'species so defined have permanent and definite differences.' A definition of Organization, or of any other term, which was not employed to express some principle, would be of no value.

The establishment, therefore, of a right Definition of a Term may be a useful step in the Explication of our Conceptions; but this will be the case *then* only when we have under our consideration some Proposition in which the Term is employed. For then the question really is, how the Conception shall be understood and defined in order that the Proposition may be true.

8. The establishment of a Proposition requires an attention to observed Facts, and can never be rightly derived from our Conceptions alone. We must hereafter consider the necessity which exists that the Facts should be rightly bound together, as well as that our Conceptions should be clearly employed, in order to lead us to real knowledge. But we may observe here that, in such cases at least as we are now considering, the two processes are co-ordinate. To unfold our Conceptions by the means of Definitions, has never been serviceable to science, except when it has been associated with an immediate *use* of the Definitions. The endeavour to define a Uniform Force was combined with the assertion that 'gravity is a uniform force:' the attempt to define Accelerating Force was immediately followed by the doctrine that 'accelerating forces may be compounded:' the process of defining Momentum was connected with the principle that 'momenta gained and lost are equal:' naturalists would have given in vain the Definition of Species which we have quoted, if they had not also given the 'characters' of species so separated. Definition and Proposition are the two handles of the instrument by which we apprehend truth; the former is of no use without the latter. Definition may be the best mode of explaining our Conception, but that which alone makes it worth while to explain it in any mode, is the opportunity of using it in the expression of Truth. When a

Definition is propounded to us as a useful step in knowledge, we are always entitled to ask what Principle it serves to enunciate. If there be no answer to this inquiry, we define and give clearness to our conceptions in vain. While we labour at such a task, we do but light up a vacant room;—we sharpen a knife with which we have nothing to cut;—we take exact aim, while we load our artillery with blank cartridge;—we apply strict rules of grammar to sentences which have no meaning.

If, on the other hand, we have under our consideration a proposition probably established, every step which we can make in giving distinctness and exactness to the Terms which this proposition involves, is an important step towards scientific truth. In such cases, any improvement in our Definition is a real advance in the explication of our Conception. The clearness of our Expressions casts a light upon the Ideas which we contemplate and convey to others.

9. But though *Definition* may be subservient to a right explication of our conceptions, it is *not essential* to that process. It is absolutely necessary to every advance in our knowledge, that those by whom such advances are made should possess clearly the conceptions which they employ: but it is by no means necessary that they should unfold these conceptions in the words of a formal Definition. It is easily seen, by examining the course of Galileo's discoveries, that he had a distinct conception of the *Moving Force* which urges bodies downwards upon an inclined plane, while he still hesitated whether to call it *Momentum, Energy, Impetus,* or *Force,* and did not venture to offer a Definition of the thing which was the subject of his thoughts. The Conception of *Polarization* was clear in the minds of many optical speculators, from the time of Huyghens and Newton to that of Young and Fresnel. This Conception we have defined to be 'Opposite properties depending upon opposite positions;' but this notion was, by the discoverers, though constantly assumed and expressed by means of superfluous hypotheses, never clothed in definite language. And in the mean time, it was the custom, among subordinate writers on the same subjects, to say, that the term *Polarization* had no definite meaning, and was merely an expression of our ignorance. The Definition which was offered by Haüy and others of a *Mineralogical Species;*—'The same elements combined in the same proportions,

with the same fundamental form;'—was false, inasmuch as it was incapable of being rigorously applied to any one case; but this defect did not prevent the philosophers who propounded such a Definition from making many valuable additions to mineralogical knowledge, in the way of identifying some species and distinguishing others. The right Conception which they possessed in their minds prevented their being misled by their own very erroneous Definition. The want of any precise Definitions of *Strata,* and *Formations* and *Epochs,* among geologists, has not prevented the discussions which they have carried on upon such subjects from being highly serviceable in the promotion of geological knowledge. For however much the apparent vagueness of these terms might leave their arguments upon to cavil, there was a general understanding prevalent among the most intelligent cultivators of the science, as to what was meant in such expressions; and this common understanding sufficed to determine what evidence should be considered conclusive and what inconclusive, in these inquiries. And thus the distinctness of Conception, which is a real requisite of scientific progress, existed in the minds of the inquirers, although Definitions, which are a partial and accidental evidence of this distinctness, had not yet been hit upon. The Idea had been developed in men's minds, although a clothing of words had not been contrived for it, nor, perhaps, the necessity of such a vehicle felt: and thus that essential condition of the progress of knowledge, of which we are here speaking, existed; while it was left to the succeeding speculators to put this unwritten Rule in the form of a verbal Statute.

10. Men are often prone to consider it as a thoughtless *omission* of an essential circumstance, and as a *neglect* which involves some blame, when knowledge thus assumes a form in which Definitions, or rather Conceptions, are implied but are not expressed. But in such a judgment, they assume *that* to be a matter of choice requiring attention only, which is in fact as difficult and precarious as any other portion of the task of discovery. To *define,* so that our Definition shall have any scientific value, requires no small portion of that sagacity by which truth is detected. As we have already said, Definitions and Propositions are co-ordinate in their use and in their origin. In many cases, perhaps in most, the Proposition which contains a scientific truth, is apprehended with confidence, but with some

vagueness and vacillation, before it is put in a positive, distinct, and definite form. It is thus known to be true, before it can be enunciated in terms each of which is rigorously defined. The business of Definition is part of the business of discovery. When it has been clearly seen what ought to be our Definition, it must be pretty well known what truth we have to state. The Definition, as well as the discovery, supposes a decided step in our knowledge to have been made. The writers on Logic in the middle ages, made *Definition* the last stage in the progress of knowledge; and in this arrangement at least, the history of science, and the philosophy derived from the history, confirm their speculative views. If the Explication of our Conceptions ever assume the form of a Definition, this will come to pass, not as an arbitrary process, or as a matter of course, but as the mark of one of those happy efforts of sagacity to which all the successive advances of our knowledge are owing.

***Sect. III. Use of Axioms***

11. Our Conceptions, then, even when they become so clear as the progress of knowledge requires, are not adequately expressed, or necessarily expressed at all, by means of Definitions. We may ask, then, whether there is any *other mode* of expression in which we may look for the evidence and exposition of that peculiar exactness of thought which the formation of Science demands. And in answer to this inquiry, we may refer to the discussions respecting many of the Fundamental Ideas of the sciences contained in our *History* of such Ideas. It has there been seen that these Ideas involve many elementary truths which enter into the texture of our knowledge, introducing into it connexions and relations of the most important kind, although these elementary truths cannot be deduced from any verbal definition of the idea. It has been seen that these elementary truths may often be enunciated by means of *Axioms,* stated in addition to, or in preference to, Definitions. For example, the Idea of Cause, which forms the basis of the science of Mechanics, makes its appearance in our elementary mechanical reasonings, not as a Definition, but by means of the Axioms that 'Causes are measured by their effects,' and that 'Reaction is equal and opposite to action.' Such axioms, tacitly assumed or occasionally stated, as maxims of acknowledged validity,

belong to all the Ideas which form the foundations of the sciences, and are constantly employed in the reasoning and speculations of those who think clearly on such subjects. It may often be a task of some difficulty to detect and enunciate in words the Principles which are thus, perhaps silently and unconsciously, taken for granted by those who have a share in the establishment of scientific truth: but inasmuch as these Principles are an essential element in our knowledge, it is very important to our present purpose to separate them from the associated materials, and to trace them to their origin. This accordingly I attempted to do, with regard to a considerable number of the most prominent of such Ideas, in the *History*. The reader will there find many of these Ideas resolved into Axioms and Principles by means of which their effect upon the elementary reasonings of the various sciences may be expressed. That Work is intended to form, in some measure, a representation of the Ideal Side of our physical knowledge;—a Table of those contents of our Conceptions which are not received directly from facts;—an exhibition of Rules to which we know that truth must conform.

### *Sect. IV. Clear and appropriate Ideas*

12. In order, however, that we may see the necessary cogency of these rules, we must possess, clearly and steadily, the Ideas from which the rules flow. In order to perceive the necessary relations of the Circles of the Sphere, we must possess clearly the Idea of Solid Space:—in order that we may see the demonstration of the composition of forces, we must have the Idea of Cause moulded into a distinct Conception of Statical Force. This is that *Clearness of Ideas* which we stipulate for in any one's mind, as the first essential condition of his making any new step in the discovery of truth. And we now see what answer we are able to give, if we are asked for a Criterion of this Clearness of Idea. The Criterion is, that the person shall *see* the necessity of the Axioms belonging to each Idea;—shall accept them in such a manner as to perceive the cogency of the reasonings founded upon them. Thus, a person has a clear Idea of Space who follows the reasonings of geometry and fully apprehends their conclusiveness. The Explication of Conceptions, which we are speaking of as an essential part of real knowledge, is the process by

which we bring the Clearness of our Ideas to bear upon the Formation of our knowledge. And this is done, as we have now seen, not always, nor generally, nor principally, by laying down a Definition of the Conception; but by acquiring such a possession of it in our minds as enables, indeed compels us, to admit, along with the Conception, all the Axioms and Principles which it necessarily implies, and by which it produces its effect upon our reasonings.

13. But in order that we may make any real advance in the discovery of truth, our Ideas must not only be clear, they must also be *appropriate*. Each science has for its basis a different class of Ideas; and the steps which constitute the progress of one science can never be made by employing the Ideas of another kind of science. No genuine advance could ever be obtained in Mechanics by applying to the subject the Ideas of Space and Time merely:—no advance in Chemistry, by the use of mere Mechanical Conceptions:—no discovery in Physiology, by referring facts to mere Chemical and Mechanical Principles. Mechanics must involve the Conception of *Force;*—Chemistry, the Conception of *Elementary Composition;*—Physiology, the Conception of *Vital Powers*. Each science must advance by means of its *appropriate* Conceptions. Each has its own field, which extends as far as its principles can be applied. I have already noted the separation of several of these fields by the divisions of the Books of the *History* of Ideas. The Mechanical, the Secondary Mechanical, the Chemical, the Classificatory, the Biological Sciences form so many great Provinces in the Kingdom of knowledge, each in a great measure possessing its own peculiar fundamental principles. Every attempt to build up a new science by the application of principles which belong to an old one, will lead to frivolous and barren speculations.

This truth has been exemplified in all the instances in which subtle speculative men have failed in their attempts to frame new sciences, and especially in the essays of the ancient schools of philosophy in Greece, as has already been stated in the History of Science. Aristotle and his followers endeavoured in vain to account for the mechanical relation of forces in the lever by applying the *inappropriate* geometrical conceptions of the properties of the circle:—they speculated to no purpose about the elementary composition of bodies, because they assumed the *inappropriate* conception of *likeness* between the elements and the compound, instead of the genuine notion of elements merely *determining* the qualities of the compound. And in like manner, in modern times, we have seen, in the history of the fundamental ideas of the physiological sciences, how all the *inappropriate* mechanical and chemical and other ideas which were applied in succession to the subject failed in bringing into view any genuine physiological truth.

14. That the real cause of the failure in the instances above mentioned lay in the *Conceptions*, is plain. It was not ignorance of the facts which in these cases prevented the discovery of the truth. Aristotle was as well acquainted with the fact of the proportion of the weights which balance on a Lever as Archimedes was, although Archimedes alone gave the true mechanical reason for the proportion.

With regard to the doctrine of the Four Elements indeed, the inapplicability of the conception of composition of qualities, required, perhaps, to be proved by some reference to facts. But this conception was devised at first, and accepted by succeeding times, in a blind and gratuitous manner, which could hardly have happened if men had been awake to the necessary condition of our knowledge;—that the conceptions which we introduce into our doctrines are not arbitrary or accidental notions, but certain peculiar modes of apprehension strictly determined by the subject of our speculations.

15. It may, however, be said that this injunction that we are to employ *appropriate* Conceptions only in the formation of our knowledge, cannot be of practical use, because we can only determine what Ideas *are* appropriate, by finding that they truly combine the facts. And this is to a certain extent true. Scientific discovery must ever depend upon some happy thought, of which we cannot trace the origin;—some fortunate cast of intellect, rising above all rules. No maxims can be given which inevitably lead to discovery. No precepts will elevate a man of ordinary endowments to the level of a man of genius: nor will an inquirer of truly inventive mind need to come to the teacher of inductive philosophy to learn how to exercise the faculties which nature has given him. Such persons as Kepler or Fresnel, or Brewster, will have their powers of discovering truth little aug-

mented by any injunctions respecting Distinct and Appropriate Ideas; and such men may very naturally question the utility of rules altogether.

16. But yet the opinions which such persons may entertain, will not lead us to doubt concerning the value of the attempts to analyse and methodize the process of discovery. Who would attend to Kepler if he had maintained that the speculations of Francis Bacon were worthless? Notwithstanding what has been said, we may venture to assert that the Maxim which points out the necessity of Ideas appropriate as well as clear, for the purpose of discovering truth is not without its use. It may, at least, have a value as a caution or prohibition, and may thus turn us away from labours certain to be fruitless. We have already seen, in the *History* of Ideas, that this maxim, if duly attended to, would have at once condemned, as wrongly directed, the speculations of physiologists of the mathematical, mechanical, chemical, and vital-fluid schools; since the Ideas which the teachers of these schools introduce, cannot suffice for the purposes of physiology, which seeks truths respecting the vital powers. Again, it is clear from similar considerations that no definition of a mineralogical species by chemical characters alone can answer the end of science, since we seek to make mineralogy, not an analytical but a classificatory science. Even before the appropriate conception is matured in men's minds so that they see clearly what it is, they may still have light enough to see what it is not.

17. Another result of this view of the necessity of appropriate Ideas, combined with a survey of the history of science is, that though for the most part, as we shall see, the progress of science consists in accumulating and combining Facts rather than in debating concerning Definitions; there are still certain periods when the *discussion* of Definitions may be the most useful mode of cultivating some special branch of science. This discussion is of course always to be conducted by the light of facts; and, as has already been said, along with the settlement of every good Definition will occur the corresponding establishment of some Proposition. But still at particular periods, the want of a Definition, or of the clear conceptions which Definition supposes, may be peculiarly felt. A good and tenable Definition of *Species* in Mineralogy would at present be perhaps the most im-

portant step which the science could make. A just conception of the nature of *Life,* (and if expressed by means of a Definition, so much the better,) can hardly fail to give its possessor an immense advantage in the speculations which now come under the considerations of physiologists. And controversies respecting Definitions, in these cases, and such as these, may be very far from idle and unprofitable.

Thus the knowledge that Clear and Appropriate Ideas are requisite for discovery, although it does not lead to any very precise precepts, or supersede the value of natural sagacity and inventiveness, may still be of use to us in our pursuit after truth. It may show us what course of research is, in each stage of science, recommended by the general analogy of the history of knowledge; and it may both save us from hopeless and barren paths of speculation, and make us advance with more courage and confidence, to know that we are looking for discoveries in the manner in which they have always hitherto been made.

#### *Sect. V. Accidental Discoveries*

18. Another consequence follows from the views presented in this Chapter, and it is the last I shall at present mention. *No scientific discovery* can, with any justice, be considered *due to accident.* In whatever manner facts may be presented to the notice of a discoverer, they can never become the materials of exact knowledge, except they find his mind already provided with precise and suitable conceptions by which they may be analysed and connected. Indeed, as we have already seen, facts cannot be observed as Facts, except in virtue of the Conceptions which the observer himself unconsciously supplies; and they are not Facts of Observation for any purpose of Discovery, except these familiar and unconscious acts of thought be themselves of a just and precise kind. But supposing the Facts to be adequately observed, they can never be combined into any new Truth, except by means of some new Conceptions, clear and appropriate, such as I have endeavoured to characterize. When the observer's mind is prepared with such instruments, a very few facts, or it may be a single one, may bring the process of discovery into action. But in such cases, this previous condition of the intellect, and not the single fact, is really the main and peculiar cause of the success. The

fact is merely the occasion by which the engine of discovery is brought into play sooner or later. It is, as I have elsewhere said, only the spark which discharges a gun already loaded and pointed; and there is little propriety in speaking of such an accident as the cause why the bullet hits the mark. If it were true that the fall of an apple was the occasion of Newton's pursuing the train of thought which led to the doctrine of universal gravitation, the habits and constitution of Newton's intellect, and not the apple, were the real source of this great event in the progress of knowledge. The common love of the marvellous, and the vulgar desire to bring down the greatest achievements of genius to our own level, may lead men to ascribe such results to any casual circumstances which accompany them; but no one who fairly considers the real nature of great discoveries, and the intellectual processes which they involve, can seriously hold the opinion of their being the effect of accident.

19. Such accidents never happen to common men. Thousands of men, even of the most inquiring and speculative men, had seen bodies fall; but who, except Newton, ever followed the accident to such consequences? And in fact, how little of his train of thought was contained in, or even directly suggested by, the fall of the apple! If the apple fall, said the discoverer, 'why should not the moon, the planets, the satellites, fall?' But how much previous thought,—what a steady conception of the universality of the laws of motion gathered from other sources,—were requisite, that the inquirer should see any connexion in these cases! Was it by accident that he saw in the apple an image of the moon, and of every body in the solar system?

20. The same observations may be made with regard to the other cases which are sometimes adduced as examples of accidental discovery. It has been said, 'By the accidental placing of a rhomb of calcareous spar upon a book or line Bartholinus discovered the property of the *Double Refraction* of light.' But Bartholinus could have seen no such consequence in the accident if he had not previously had a clear conception of *single refraction*. A lady, in describing an optical experiment which had been shown her, said of her teacher, 'He told me to *increase and diminish the angle of refraction*, and at last I found that he only meant me to move my head up and down.' At any rate, till the lady had acquired the notions which the technical

terms convey, she could not have made Bartholinus's discovery by means of his accident. 'By accidentally combining two rhombs in different positions,' it is added, 'Huyghens discovered the *Polarization* of Light.' Supposing that this experiment had been made without design, what Huyghens really observed was, that the images appeared and disappeared alternately as he turned one of the rhombs round. But was it an easy or an obvious business to analyze this curious alternation into the circumstances of the rays of light having *sides*, as Newton expressed it, and into the additional hypotheses which are implied in the term 'polarization'? Those will be able to answer this question, who have found how far from easy it is to understand clearly what is meant by 'polarization' in this case, now that the property is fully established. Huyghens's success depended on his clearness of thought, for this enabled him to perform the intellectual analysis, which never would have occurred to most men, however often they had 'accidentally combined two rhombs in different positions.' 'By accidentally looking through a prism of the same substance, and turning it round, Malus discovered the polarization of light by reflection.' Malus saw that, in some positions of the prism, the light reflected from the windows of the Louvre thus seen through the prism, became dim. A common man would have supposed this dimness the result of accident; but Malus's mind was differently constituted and disciplined. He considered the position of the window, and of the prism; repeated the experiment over and over; and in virtue of the eminently distinct conceptions of space which he possessed, resolved the phenomena into its geometrical conditions. A believer in accident would not have sought them; a person of less clear ideas would not have found them. A person must have a strange confidence in the virtue of chance, and the worthlessness of intellect, who can say that 'in all these fundamental discoveries appropriate ideas had no share,' and that the discoveries 'might have been made by the most ordinary observers.'

21. I have now, I trust, shown in various ways, how the *Explication of Conceptions*, including in this term their clear development from Fundamental Ideas in the discoverer's mind, as well as their precise expression in the form of Definitions or Axioms, when that can be done, is an essential part in the establishment of all

exact and general physical truths. In doing this, I have endeavoured to explain in what sense the possession of clear and appropriate ideas is a main requisite for every step in scientific discovery. That it is far from being the only step, I shall soon have to show; and if any obscurity remain on the subject treated of in the present chapter, it will, I hope, be removed when we have examined the other elements which enter into the constitution of our knowledge.

## CHAPTER III
## OF FACTS AS THE MATERIALS OF SCIENCE

***Aphorism IV***

*Facts are the materials of science, but all Facts involve Ideas. Since, in observing Facts, we cannot exclude Ideas, we must, for the purposes of science, take care that the Ideas are clear and rigorously applied.*

***Aphorism V***

*The last Aphorism leads to such Rules as the following:—That Facts, for the purposes of material science, must involve Conceptions of the Intellect only, and not Emotions:—That Facts must be observed with reference to our most exact conceptions, Number, Place, Figure, Motion:—That they must also be observed with reference to any other exact conceptions which the phenomena suggest, as Force, in mechanical phenomena, Concord, in musical.*

***Aphorism VI***

*The resolution of complex Facts into precise and measured partial Facts, we call the* Decomposition of Facts. *This process is requisite for the progress of science, but does not necessarily lead to progress.*

1. We have now to examine how Science is built up by the combination of Facts. In doing this, we suppose that we have already attained a supply of definite and certain Facts, free from obscurity and doubt. We must, therefore, first consider under what conditions Facts can assume this character.

When we inquire what Facts are to be made the materials of Science, perhaps the answer which we should most commonly receive would be, that they must be *True Facts,* as distinguished from any mere inferences or opinions of our own. We should probably be told that we must be careful in such a case to consider as Facts, only what we really observe;—that we must assert only what we see; and believe nothing except upon the testimony of our senses.

But such maxims are far from being easy to apply, as a little examination will convince us.

2. It has been explained, in preceding works, that all perception of external objects and occurrences involves an active as well as a passive process of the mind;—includes not only Sensations, but also Ideas by which Sensations are bound together, and have a unity given to them. From this it follows, that there is a difficulty in separating in our perceptions what we receive from without, and what we ourselves contribute from within;—what we perceive, and what we infer. In many cases, this difficulty is obvious to all: as, for example, when we witness the performances of a juggler or a ventriloquist. In these instances, we imagine ourselves to see and to hear what certainly we do not see and supply interruptions and infer connexions; and by giving us fallacious hear. The performer takes advantage of the habits by which our minds indications, he leads us to perceive as an actual fact, what does not happen at all. In these cases, it is evident that we ourselves assist in making the fact; for we make one which does not really exist. In other cases, though the fact which we perceive be true, we can easily see that a large portion of the perception is our own act; as when, from the sight of a bird of prey we infer a carcase, or when we read a half-obliterated inscription. In the latter case, the mind supplies the meaning, and perhaps half the letters; yet we do not hesitate to say that we actually *read* the inscription. Thus, in many cases, our own inferences and interpretations enter into our facts. But this happens in many instances in which it is at first sight less obvious. When any one has seen an oak-tree blown down by a strong gust of wind, he does not think of the occurrence any otherwise than as a *Fact* of which he is assured by his senses. Yet by what sense does he perceive the Force which he thus supposes the wind to exert? By what sense does he distinguish an Oak-tree from all other trees? It is clear upon reflexion, that in such a case, his own mind supplies the conception of extraneous impulse and pressure, by which he thus interprets the motions

observed, and the distinction of different kinds of trees, according to which he thus names the one under his notice. The Idea of Force, and the idea of definite Resemblances and Differences, are thus combined with the impressions on our senses, and form an indistinguished portion of that which we consider as the Fact. And it is evident that we can in no other way perceive Force, than by seeing motion; and cannot give a Name to any object, without not only seeing a difference of single objects, but supposing a difference of classes of objects. When we speak as if we saw impulse and attraction, things and classes, we really see only objects of various forms and colours, more or less numerous, variously combined. But do we really perceive so much as this? When we see the form, the size, the number, the motion of objects, are these really mere impressions on our senses, unmodified by any contribution or operation of the mind itself? A very little attention will suffice to convince us that this is not the case. When we see a windmill turning, it may happen, as we have elsewhere noticed,[1] that we mistake the direction in which the sails turn: when we look at certain diagrams, they may appear either convex or concave: when we see the moon first in the horizon and afterwards high up in the sky, we judge her to be much larger in the former than in the latter position, although to the eye she subtends the same angle. And in these cases and the like, it has been seen that the errour and confusion which we thus incur arise from the mixture of acts of the mind itself with impressions on the senses. But such acts are, as we have also seen, *inseparable* portions of the process of perception. A certain activity of the mind is involved, not only in seeing objects erroneously, but in seeing them at all. With regard to solid objects, this is generally acknowledged. When we seem to see an edifice occupying space in all dimensions, we really see only a representation of it as it appears referred by perspective to a surface. The inference of the solid form is an operation of our own, alike when we look at a reality and when we look at a picture. But we may go further. Is plane Figure really a mere Sensation? If we look at a decagon, do we see at once that it has ten sides or is it not necessary for us to count them: and is not counting an act of the mind? All objects are seen in space; all objects are seen as one or many: but are not the Idea of Space and the Idea of Number requisite in order that we may thus apprehend what we see? That these

Ideas of Space and Number involve a connexion derived from the mind, and not from the senses, appears, as we have already seen, from this, that those Ideas afford us the materials of universal and necessary truths:—such truths as the senses cannot possibly supply. And thus, even the perception of such facts as the size, shape, and number of objects, cannot be said to be impressions of sense, distinct from all acts of mind, and cannot be expected to be free from errour on the ground of their being mere observed Facts.

Thus the difficulty which we have been illustrating, of distinguishing Facts from inferences and from interpretations of facts, is not only great, but amounts to an impossibility. The separation at which we aimed in the outset of this discussion, and which was supposed to be necessary in order to obtain a firm groundwork for science, is found to be unattainable. We cannot obtain a sure basis of Facts, by rejecting all inferences and judgments of our own, for such inferences and judgments form an unavoidable element in all Facts. We cannot exclude our Ideas from our Perceptions, for our Perceptions involve our Ideas.

3. But still, it cannot be doubted that in selecting the Facts which are to form the foundation of Science, we must reduce them to their most simple and certain form; and must reject everything from which doubt or errour may arise. Now since this, it appears, cannot be done, by rejecting the Ideas which all Facts involve, in what manner are we to conform to the obvious maxim, that the Facts which form the basis of Science must be perfectly definite and certain?

The analysis of facts into Ideas and Sensations, which we have so often referred to, suggests the answer to this inquiry. We are not able, nor need we endeavour, to exclude Ideas from our Facts; but we may be able to discern, with perfect distinctness, the Ideas which we include. We cannot observe any phenomena without applying to them such Ideas as Space and Number, Cause and Resemblance, and usually, several others; but we may avoid applying these Ideas in a wavering or obscure manner, and confounding Ideas with one another. We cannot read any of the inscriptions which nature presents to us, without interpreting them by means of some language which we ourselves are accustomed to speak; but we may make it our business to acquaint ourselves perfectly with the language which we thus employ, and to interpret it according to the rigorous rules of grammar and analogy.

This maxim, that when Facts are employed as the basis of Science, we must distinguish clearly the Ideas which they involve, and must apply these in a distinct and rigorous manner, will be found to be a more precise guide than we might perhaps at first expect. We may notice one or two Rules which flow from it.

4. In the first place, Facts, when used as the materials of physical Science, must be *referred to Conceptions of the Intellect only,* all emotions of fear, admiration, and the like, being rejected or subdued. Thus, the observations of phenomena which are related as portents and prodigies, striking terrour and boding evil, are of no value for purposes of science. The tales of armies seen warring in the sky, the sound of arms heard from the clouds, fiery dragons, chariots, swords seen in the air, may refer to meterorological phenomena; but the records of phenomena observed in the state of mind which these descriptions imply can be of no scientific value. We cannot make the poets our observers.

> Armorum sonitum toto Germania cœlo
> Audiit; insolitis tremuerunt motibus Alpes.
> Vox quoque per lucos vulgo exaudita silentes
> Ingens, et simulacra modis pallentia miris
> Visa sub obscurum noctis: pecudesque locutæ.

The mixture of fancy and emotion with the observation of facts has often disfigured them to an extent which is too familiar to all to need illustration. We have an example of this result, in the manner in which Comets are described in the treatises of the middle ages. In such works, these bodies are regularly distributed into several classes, accordingly as they assume the form of a sword, of a spear, of a cross, and so on. When such resemblances had become matters of interest, the impressions of the senses were governed, not by the rigorous conceptions of form and colour, but by these assumed images; and under these circumstances, we can attach little value to the statement of what was seen.

In all such phenomena, the reference of the objects to the exact Ideas of Space, Number, Position, Motion, and the like, is the first step of Science: and accordingly, this reference was established at an early period in those sciences which made an early progress, as, for instance, Astronomy. Yet even in astronomy there appears to have been a period when the predominant conceptions of men in regarding the heavens and the stars pointed to mythical story and supernatural influence rather than to mere relations of space, time, and motion: and of this primeval condition of those who gazed at the stars, we seem to have remnants in the Constellations, in the mythological Names of the Planets, and in the early prevalence of Astrology. It was only at a later period, when men had begun to measure the places, or at least to count the revolutions of the stars, that Astronomy had its birth.

5. And thus we are led to another Rule:—that in collecting Facts which are to be made the basis of Science, the Facts are to be observed, as far as possible, *with reference to place, figure, number, motion,* and the like Conceptions; which, depending upon the Ideas of Space and Time, are the most universal, exact, and simple of our conceptions. It was by early attention to these relations in the case of the heavenly bodies, that the ancients formed the science of Astronomy: it was by not making precise observations of this kind in the case of terrestrial bodies, that they failed in framing a science of the Mechanics of Motion. They succeeded in Optics as far as they made observations of this nature; but when they ceased to trace the geometrical paths of rays in the actual experiment, they ceased to go forwards in the knowledge of this subject.

6. But we may state a further Rule:—that though these relations of Time and Space are highly important in almost all Facts, we are not to confine ourselves to these: but are to consider the phenomena *with reference to other Conceptions also:* it being always understood that these conceptions are to be made as exact and rigorous as those of geometry and number. Thus the science of Harmonics arose from considering sounds with reference to *Concords* and *Discords;* the science of Mechanics arose from not only observing motions as they take place in Time and Space, but further, referring them to *Force* as their *Cause.* And in like manner, other sciences depend upon other Ideas, which, as I have endeavoured to show, are not less fundamental than those of Time and Space; and like them, capable of leading to rigorous consequences.

7. Thus the Facts which we assume as the basis of Science are to be freed from all the mists which imagination and passion throw round them; and to be separated into those elementary Facts which exhibit simple and evident relations of Time, or Space, or Cause, or some other Ideas equally clear. We resolve the complex appearances which nature offers to us, and the mixed and manifold modes of looking at these appearances which rise in our thoughts, into limited, definite, and clearly-understood portions. This process we may term the *Decomposition of Facts.* It is the beginning of exact knowledge, —the first step in the formation of all Science. This Decomposition of Facts into Elementary Facts, clearly understood and surely ascertained, must precede all discovery of the laws of nature.

8. But though this step is necessary, it is not infallibly sufficient. It by no means follows that when we have thus decomposed Facts into Elementary Truths of observation, we shall soon be able to combine these, so as to obtain Truths of a higher and more speculative kind. We have examples which show us how far this is from being a necessary consequence of the former step. Observations of the weather, made and recorded for many years, have not led to any general truths, forming a science of Meteorology: and although great numerical precision has been given to such observation by means of barometers, thermometers, and other instruments, still, no general laws regulating the cycles of change of such phenomena have yet been discovered. In like manner the faces of crystals, and the sides of the polygons which these crystals form, were counted, and thus numerical facts were obtained, perfectly true and definite, but still of no value for purposes of science. And when it was discovered what Element of the form of crystals it was important to observe and measure, namely, the Angle made by two faces with each other, this discovery was a step of a higher order, and did not belong to that department, of mere exact observation of manifest Facts, with which we are here concerned.

9. When the Complex Facts which nature offers to us are thus decomposed into Simple Facts, the decomposition, in general, leads to the introduction of *Terms* and Phrases, more or less technical, by which these Simple Facts are described. When Astronomy was thus made a science of measurement, the things measured were soon described as *Hours,* and *Days,* and *Cycles, Altitude* and *Declination, Phases* and *Aspects.* In the same manner, in Music, the concords had

names assigned them, as *Diapente, Diatessaraon, Diapason;* in studying Optics, the *Rays* of light were spoken of as having their course altered by *Reflexion* and *Refraction;* and when useful observations began to be made in Mechanics, the observers spoke of *Force, Pressure, Momentum, Inertia,* and the like.

10. When we take phenomena in which the leading Idea is Resemblance, and resolve them into precise component Facts, we obtain some kind of Classification; as, for instance, when we lay down certain Rules by which particular trees, or particular animals are to be known. This is the earliest form of Natural History; and the Classification which it involves is that which corresponds, nearly or exactly, with the usual Names of the objects thus classified.

11. Thus the first attempts to render observation certain and exact, lead to a decomposition of the obvious facts into Elementary Facts, connected by the Ideas of Space, Time, Number, Cause, Likeness, and others: and into a Classification of the Simple Facts; a classification more or less just, and marked by Names either common or technical. Elementary Facts, and Individual Objects thus observed and classified, form the materials of Science; and any improvement in Classification or Nomenclature, or any discovery of a Connexion among the materials thus accumulated, leads us fairly within the precincts of Science. We must now, therefore, consider the manner in which Science is built up of such materials;—the process by which they are brought into their places, the texture of the bond which unites and cements them.

## CHAPTER IV
## OF THE COLLIGATION OF FACTS

### *Aphorism VII*

*Science begins with* common *observation of facts; but even at this stage, requires that the observations be precise. Hence the sciences which depend upon space and number were the earliest formed. After common observation, come Scientific* Observation *and* Experiment.

### *Aphorism VIII*

*The Conceptions by which Facts are bound together, are suggested by the sagacity of discoverers. This sagacity cannot be taught.*

*It commonly succeeds by guessing; and this success seems to consist in framing several* tentative hypotheses *and selecting the right one. But a supply of appropriate hypotheses cannot be constructed by rule, nor without inventive talent.*

*Aphorism IX*

*The truth of tentative hypotheses must be tested by their application to facts. The discoverer must be ready, carefully to try his hypotheses in this manner, and to reject them if they will not bear the test, in spite of indolence and vanity.*

1. Facts such as the last Chapter speaks of are, by means of such Conceptions as are described in the preceding Chapter, bound together so as to give rise to those general Propositions of which Science consists. Thus the Facts that the planets revolve about the sun in certain periodic times and at certain distances, are included and connected in Kepler's Law, by means of such Conceptions as the *squares of numbers,* the *cubes of distances,* and the *proportionality* of these quantities. Again the existence of this proportion in the motions of any two planets, forms a set of Facts which may all be combined by means of the Conception of a certain *central accelerating force,* as was proved by Newton. The whole of our physical knowledge consists in the establishment of such propositions; and in all such cases, Facts are bound together by the aid of suitable Conceptions. This part of the formation of our knowledge I have called the *Colligation of Facts* and we may apply this term to every case in which, by an act of the intellect, we establish a precise connexion among the phenomena which are presented to our senses. The knowledge of such connexions, accumulated and systematized, is Science. On the steps by which science is thus collected from phenomena we shall proceed now to make a few remarks.

2. Science begins with *Common* Observation of facts, in which we are not conscious of any peculiar discipline or habit of thought exercised in observing. Thus the common perceptions of the appearances and recurrences of the celestial luminaries, were the first steps of Astronomy: the obvious cases in which bodies fall or are supported, were the beginning of Mechanics; the familiar aspects of visible things, were the origin of Optics; the usual distinctions of well-known

plants, first gave rise to Botany. Facts belonging to such parts of our knowledge are noticed by us, and accumulated in our memories, in the common course of our habits, almost without our being aware that we are observing and collecting facts. Yet such facts may lead to many scientific truths; for instance, in the first stages of Astronomy (as we have shown in the *History*) such facts led to Methods in Intercalation and Rules of the Recurrence of Eclipses. In succeeding stages of science, more especial attention and preparation on the part of the observer, and a selection of certain *kinds* of facts, becomes necessary; but there is an early period in the progress of knowledge at which man is a physical philosopher, without seeking to be so, or being aware that he is so.

3. But in all states of the progress, even in that early one of which we have just spoken, it is necessary, in order that the facts may be fit materials of any knowledge, that they should be decomposed into Elementary Facts, and that these should be observed with precision. Thus, in the first infancy of astronomy, the recurrence of phases of the moon, of places of the sun's rising and setting, of planets, of eclipses, was observed to take place at intervals of certain definite numbers of days, and in a certain exact order; and thus it was, that the observations became portions of astronomical science. In other cases, although the facts were equally numerous, and their general aspect equally familiar, they led to no science, because their exact circumstances were not apprehended. A vague and loose mode of looking at facts very easily observable, left men for a long time under the belief that a body, ten times as heavy as another, falls ten times as fast; —that objects immersed in water are always magnified, without regard to the form of the surface;—that the magnet exerts an irresistible force;—that crystal is always found associated with ice;—and the like. These and many others are examples how blind and careless men can be, even in observation of the plainest and commonest appearances; and they show us that the mere faculties of perception, although constantly exercised upon innumerable objects, may long fail in leading to an exact knowledge.

If we further inquire what was the favourable condition through which some special classes of facts were, from the first, fitted to become portions of science, we shall find it to have been principally this;—that these facts were considered with reference to the Ideas of Time, Number, and Space, which are Ideas possessing peculiar definiteness and precision; so that with regard to them, confusion and indistinctness are hardly possible. The interval from new moon to new moon was always a particular number of days: the sun in his yearly course rose and set near to a known succession of distant objects: the moon's path passed among the stars in a certain order:—these are observations in which mistake and obscurity are not likely to occur, if the smallest degree of attention is bestowed upon the task. To count a number is, from the first opening of man's mental faculties, an operation which no science can render more precise. The relations of space are nearest to those of number in obvious and universal evidence. Sciences depending upon these Ideas arise with the first dawn of intellectual civilization. But few of the other Ideas which man employs in the acquisition of knowledge possess this clearness in their common use. The Idea of *Resemblance* may be noticed, as coming next to those of Space and Number in original precision; and the Idea of *Cause,* in a certain vague and general mode of application, sufficient for the purposes of common life, but not for the ends of science, exercises a very extensive influence over men's thoughts. But the other Ideas on which science depends, with the Conceptions which arise out of them, are not unfolded till a much later period of intellectual progress; and therefore, except in such limited cases as I have noticed, the observations of common spectators and uncultivated nations, however numerous or varied, are of little or no effect in giving rise to Science.

5. Let us now suppose that, besides common everyday perception of facts, we turn our attention to some other occurrences and appearances, with a design of obtaining from them speculative knowledge. This process is more peculiarly called *Observation,* or, when we ourselves occasion the facts, *Experiment.* But the same remark which we have already made, still holds good here. These facts can be of no value, except they are resolved into those exact Conceptions which contain the essential circumstances of the case. They must be determined, not indeed necessarily, as has sometimes been said, 'according to Number, Weight, and Measure;' for, as we have endeavoured to show in the preceding Books,[2] there are many other Conceptions to which phenomena may be subordinated, quite different from these,

and yet not at all less definite and precise. But in order that the facts obtained by observation and experiment may be capable of being used in furtherance of our exact and solid knowledge, they must be apprehended and analysed according to some Conceptions which, applied for this purpose, give distinct and definite results, such as can be steadily taken hold of and reasoned from; that is, the facts must be referred to Clear and Appropriate Ideas, according to the manner in which we have already explained this condition of the derivation of our knowledge. The phenomena of light, when they are such as to indicate sides in the ray, must be referred to the Conception of *polarization;* the phenomena of mixture, when there is an alteration of qualities as well as quantities, must be combined by a Conception of *elementary composition.* And thus, when mere position, and number, and resemblance, will no longer answer the purpose of enabling us to connect the facts, we call in other Ideas, in such cases more efficacious, though less obvious.

6. But how are we, in these cases, to discover such Ideas, and to judge which will be efficacious, in leading to a scientific combination of our experimental data? To this question, we must in the first place answer, that the first and great instrument by which facts, so observed with a view to the formation of exact knowledge, are combined into important and permanent truths, is that peculiar Sagacity which belongs to the genius of a Discoverer; and which, while it supplies those distinct and appropriate Conceptions which lead to its success, cannot be limited by rules, or expressed in definitions. It would be difficult or impossible to describe in words the habits of thought which led Archimedes to refer the conditions of equilibrium on the Lever to the Conception of *pressure,* while Aristotle could not see in them anything more than the results of the strangeness of the properties of the circle;—or which impelled Pascal to explain by means of the Conception of the *weight of air,* the facts which his predecessors had connected by the notion of nature's horrour of a vacuum;—or which caused Vitello and Roger Bacon to refer the magnifying power of a convex lens to the bending of the rays of light towards the perpendicular by *refraction,* while others conceived the effect to result from the matter of medium, with no consideration of its form. These are what are commonly spoken of as felicitous and inexplicable strokes of inventive

talent; and such, no doubt, they are. No rules can ensure to us similar success in new cases; or can enable men who do not possess similar endowments, to make like advances in knowledge.

7. Yet still, we may do something in tracing the process by which such discoveries are made; and this it is here our business to do. We may observe that these, and the like discoveries, are not improperly described as happy *Guesses;* and that Guesses, in these as in other instances, imply various suppositions made, of which some one turns out to be the right one. We may, in such cases, conceive the discoverer as inventing and trying many conjectures, till he finds one which answers the purpose of combining the scattered facts into a single rule. The discovery of general truths from special facts is performed, commonly at least, and more commonly than at first appears, by the use of a series of Suppositions, or *Hypotheses,* which are looked at in quick succession, and of which the one which really leads to truth is rapidly detected, and when caught sight of, firmly held, verified, and followed to its consequences. In the minds of most discoverers, this process of invention, trial, and acceptance or rejection of the hypothesis, goes on so rapidly that we cannot trace it in its successive steps. But in some instances, we can do so; and we can also see that the other examples of discovery do not differ essentially from these. The same intellectual operations take place in other cases, although this often happens so instantaneously that we lose the trace of the progression. In the discoveries made by Kepler, we have a curious and memorable exhibition of this process in its details. Thanks to his communicative disposition, we know that he made nineteen hypotheses with regard to the motion of Mars, and calculated the results of each, before he established the true doctrine, that the planet's path is an ellipse. We know, in like manner, that Galileo made wrong suppositions respecting the laws of falling bodies, and Mariotte, concerning the motion of water in a siphon, before they hit upon the correct view of these cases.

8. But it has very often happened in the history of science, that the erroneous hypotheses which preceded the discovery of the truth have been made, not by the discoverer himself, but by his precursors; to whom he thus owed the service, often an important one in such cases, of exhausting the most tempting forms of errour. Thus the various

fruitless suppositions by which Kepler endeavoured to discover the law of refraction, led the way to its real detection by Snell; Kepler's numerous imaginations concerning the forces by which the celestial motions are produced,—his 'physical reasonings' as he termed them,—were a natural prelude to the truer physical reasonings of Newton. The various hypotheses by which the suspension of vapour in air had been explained, and their failure, left the field open for Dalton with his doctrine of the mechanical mixture of gases. In most cases, if we could truly analyze the operation of the thoughts of those who make, or who endeavour to make discoveries in science, we should find that many more suppositions pass through their minds than those which are expressed in words; many a possible combination of conceptions is formed and soon rejected. There is a constant invention and activity, a perpetual creating and selecting power at work, of which the last results only are exhibited to us. Trains of hypotheses are called up and pass rapidly in review; and the judgment makes its choice from the varied group.

9. It would, however, be a great mistake to suppose that the hypotheses, among which our choice thus lies, are constructed by an enumeration of obvious cases, or by a wanton alteration of relations which occur in some first hypothesis. It may, indeed, sometimes happen that the proposition which is finally established is such as may be formed, by some slight alteration, from those which are justly rejected. Thus Kepler's elliptical theory of Mars's motions, involved relations of lines and angles much of the same nature as his previous false suppositions: and the true law of refraction so much resembles those erroneous ones which Kepler tried, that we cannot help wondering how he chanced to miss it. But it more frequently happens that new truths are brought into view by the application of new Ideas, not by new modifications of old ones. The cause of the properties of the Lever was learnt, not by introducing any new *geometrical* combination of lines and circles, but by referring the properties to genuine *mechanical* Conceptions. When the Motions of the Planets were to be explained, this was done, not by merely improving the previous notions, of cycles of time, but by introducing the new conception of *epicycles* in space. The doctrine of the Four Simple Elements was expelled, not by forming any new scheme of elements which should impart, according

to new rules, their sensible qualities to their compounds, but by considering the elements of bodies as *neutralizing* each other. The Fringes of Shadows could not be explained by ascribing new properties to the single rays of light, but were reduced to law by referring them to the *interference* of several rays.

Since the true supposition is thus very frequently something altogether diverse from all the obvious conjectures and combinations, we see here how far we are from being able to reduce discovery to rule, or to give any precepts by which the want of real invention and sagacity shall be supplied. We may warn and encourage these faculties when they exist, but we cannot create them, or make great discoveries when they are absent.

10. The Conceptions which a true theory requires are very often clothed in a *Hypothesis* which connects with them several superfluous and irrelevant circumstances. Thus the Conception of the Polarization of Light was originally represented under the image of *particles* of light having their poles all turned in the same direction. The Laws of Heat may be made out perhaps most conveniently by conceiving Heat to be *Fluid*. The Attraction of Gravitation might have been successfully applied to the explanation of facts, if Newton had throughout treated Attraction as the result of an *Ether* diffused through space; a supposition which he has noticed as a possibility. The doctrine of Definite and Multiple Proportions may be conveniently expressed by the hypothesis of *Atoms*. In such cases, the Hypothesis may serve at first to facilitate the introduction of a new Conception. Thus a pervading Ether might for a time remove a difficulty, which some persons find considerable, of imagining a body to exert force at a distance. A Particle with Poles is more easily conceived than Polarization in the abstract. And if hypotheses thus employed will really explain the facts by means of a few simple assumptions, the laws so obtained may afterwards be reduced to a simpler form than that in which they were first suggested. The general laws of Heat, of Attraction, of Polarization, of Multiple Proportions, are now certain, whatever image we may form to ourselves of their ultimate causes.

11. In order, then, to discover scientific truths, suppositions consisting either of new Conceptions, or of new Combinations of old ones, are to be made, till we find one supposition which succeeds in

binding together the Facts. But how are we to find this? How is the trial to be made? What is meant by 'success' in these cases? To this we reply, that our inquiry must be, whether the Facts have the same relation in the Hypothesis which they have in reality;—whether the results or our suppositions agree with the phenomena which nature presents to us. For this purpose, we must both carefully observe the phenomena, and steadily trace the consequences of our assumptions, till we can bring the two into comparison. The Conceptions which our hypotheses involve, being derived from certain Fundamental Ideas, afford a basis of rigorous reasoning, as we have shown in the Books of the *History* of those Ideas. And the results to which this reasoning leads, will be susceptible of being verified or contradicted by observation of the facts. Thus the Epicyclical Theory of the Moon, once assumed, determined what the moon's place among the stars ought to be at any given time, and could therefore be tested by actually observing the moon's places. The doctrine that musical strings of the same length, stretched with weights of 1, 4, 9, 16, would give the musical intervals of an octave, a fifth, a fourth, in succession, could be put to the trial by any one whose ear was capable of appreciating those intervals: and the inference which follows from this doctrine by numerical reasoning,—that there must be certain imperfections in the concords of every musical scale,—could in like manner be confirmed by trying various modes of *Temperament*. In like manner all received theories in science, up to the present time, have been established by taking up some supposition, and comparing it, directly or by means of its remoter consequences, with the facts it was intended to embrace. Its agreement, under certain cautions and conditions, of which we may hereafter speak, is held to be the evidence of its truth. It answers its genuine purpose, the Colligation of Facts.

12. When we have, in any subject, succeeded in one attempt of this kind, and obtained some true Bond of Unity by which the phenomena are held together, the subject is open to further prosecution; which ulterior process may, for the most part, be conducted in a more formal and technical manner. The first great outline of the subject is drawn; and the finishing of the resemblance of nature demands a more minute pencilling, but perhaps requires less of genius in the master. In the pursuance of this task, rules and precepts may be given, and fea-

tures and leading circumstances pointed out, of which it may often be useful to the inquirer to be aware.

Before proceeding further, I shall speak of some characteristic marks which belong to such scientific processes as are now the subject of our consideration, and which may sometimes aid us in determining when the task has been rightly executed.

## CHAPTER V
## OF CERTAIN CHARACTERISTICS OF SCIENTIFIC INDUCTION

***Aphorism X***

*The process of scientific discovery is cautious and rigorous, not by abstaining from hypotheses, but by rigorously comparing hypotheses with facts, and by resolutely rejecting all which the comparison does not confirm.*

***Aphorism XI***

*Hypotheses may be useful, though involving much that is superfluous, and even erroneous: for they may supply the true bond of connexion of the facts; and the superfluity and errour may afterwards be pared away.*

***Aphorism XII***

*It is a test of true theories not only to account for, but to predict phenomena.*

***Aphorism XIII***

Induction *is a term applied to describe the* process *of a true Colligation of Facts by means of an exact and appropriate Conception.* An Induction *is also employed to denote the* proposition *which results from this process.*

***Aphorism XIV***

The Consilience of Inductions *takes place when an Induction, obtained from one class of facts, coincides with an Induction, obtained from another different class. This Consilience is a test of the truth of the Theory in which it occurs.*

***Aphorism XV***

*An Induction is not the mere* sum *of the Facts which are colligated. The Facts are not only brought together, but seen in a new point of view. A new mental Element is* superinduced; *and a peculiar constitution and discipline of mind are requisite in order to make this Induction.*

***Aphorism XVI***

*Although in Every Induction a new conception is superinduced upon the Facts; yet this once effectually done, the novelty of the conception is overlooked, and the conception is considered as a part of the fact.*

***Sect. I. Invention a part of Induction***

1. The two operations spoken of in the preceding chapters,—the Explication of the Conceptions of our own minds, and the Colligation of observed Facts by the aid of such Conceptions,—are, as we have just said, inseparably connected with each other. When united, and employed in collecting knowledge from the phenomena which the world presents to us, they constitute the mental process of *Induction;* which is usually and justly spoken of as the genuine source of all our *real general knowledge* respecting the external world. And we see, from the preceding analysis of this process into its two constituents, from what origin it derives each of its characters. It is *real,* because it arises from the combination of Real Facts, but it is *general,* because it implies the possession of General Ideas. Without the former, it would not be knowledge of the External World, without the latter, it would not be Knowledge at all. When Ideas and Facts are separated from each other, the neglect of Facts gives rise to empty speculations, idle subtleties, visionary inventions, false opinions concerning the laws of phenomena, disregard of the true aspect of nature: while the want of Ideas leaves the mind overwhelmed, bewildered, and stupified by particular sensations, with no means of connecting the past with the future, the absent with the present, the example with the rule; open

to the impression of all appearances, but capable of appropriating none. Ideas are the *Form,* facts the *Material* of our structure. Knowledge does not consist in the empty mould, or in the brute mass of matter, but in the rightly-moulded substance. Induction gathers general truths from particular facts;—and in her harvest, the corn and the reaper, the solid ears and the binding band, are alike requisite. All our knowledge of nature is obtained by Induction; the term being understood according to the explanation we have now given. And our knowledge is then most complete, then most truly deserves the name of Science, when both its elements are most perfect;—when the Ideas which have been concerned in its formation have, at every step, been clear and consistent; and when they have, at every step also, been employed in binding together real and certain Facts. Of such Induction, I have already given so many examples and illustrations in the two preceding chapters, that I need not now dwell further upon the subject.

2. Induction is familiarly spoken of as the process by which we collect a *General Proposition* from a number of *Particular Cases:* and it appears to be frequently imagined that the general proposition results from a mere juxta-position of the cases, or at most, from merely conjoining and extending them. But if we consider the process more closely, as exhibited in the cases lately spoken of, we shall perceive that this is an inadequate account of the matter. The particular facts are not merely brought together, but there is a New Element added to the combination by the very act of thought by which they are combined. There is a Conception of the mind introduced in the general proposition, which did not exist in any of the observed facts. When the Greeks, after long observing the motions of the planets, saw that these motions might be rightly considered as produced by the motion of one wheel revolving in the inside of another wheel, these Wheels were Creations of their minds, added to the Facts which they perceived by sense. And even if the wheels were no longer supposed to be material, but were reduced to mere geometrical spheres or circles, they were not the less products of the mind alone,—something additional to the facts observed. The same is the case in all other discoveries. The facts are known, but they are insulated and unconnected, till the discoverer supplies from his own stores a Principle of Connexion. The

pearls are there, but they will not hang together till some one provides the String. The distances and periods of the planets were all so many separate facts; by Kepler's Third Law they are connected into a single truth: but the Conceptions which this law involves were supplied by Kepler's mind, and without these, the facts were of no avail. The planets described ellipses round the sun, in the contemplation of others as well as of Newton; but Newton conceived the deflection from the tangent in these elliptical motions in a new light,—as the effect of a Central Force following a certain law; and then it was, that such a force was discovered truly to exist.

Thus in each inference made by Induction, there is introduced some General Conception, which is given, not by the phenomena, but by the mind. The conclusion is not contained in the premises, but includes them by the introduction of a New Generality. In order to obtain our inference, we travel beyond the cases which we have before us; we consider them as mere exemplifications of some Ideal Case in which the relations are complete and intelligible. We take a Standard, and measure the facts by it; and this Standard is constructed by us, not offered by Nature. We assert, for example, that a body left to itself will move on with unaltered velocity; not because our senses ever disclosed to us a body doing this, but because (taking this as our Ideal Case) we find that all actual cases are intelligible and explicable by means of the Conception of *Forces,* causing change and motion, and exerted by surrounding bodies. In like manner, we see bodies striking each other, and thus moving and stopping, accelerating and retarding each other: but in all this, we do not perceive by our senses that abstract quantity, *Momentum,* which is always lost by one body as it is gained by another. This Momentum is a creation of the mind, brought in among the facts, in order to convert their apparent confusion into order, their seeming chance into certainty, their perplexing variety into simplicity. This the Conception of *Momentum gained and lost* does: and in like manner, in any other case in which a truth is established by Induction, some Conception is introduced, some Idea is applied, as the means of binding together the facts, and thus producing the truth.

3. Hence in every inference by Induction, there is some Concep-

tion *superinduced* upon the Facts: and we may henceforth conceive this to be the peculiar import of the term *Induction.* I am not to be understood as asserting that the term was originally or anciently employed with this notion of its meaning; for the peculiar feature just pointed out in Induction has generally been overlooked. This appears by the accounts generally given of Induction. 'Induction,' says Aristotle,[3] 'is when by means of one extreme term[4] we infer the other extreme term to be true of the middle term.' Thus, (to take such exemplifications as belong to our subject,) from knowing that Mercury, Venus, Mars, describe ellipses about the Sun, we infer that all Planets describe ellipses about the Sun. In making this inference syllogistically, we assume that the evident proposition, 'Mercury, Venus, Mars, do what all Planets do,' may be taken *conversely,* 'All Planets do what Mercury, Venus, Mars, do.' But we may remark that, in this passage, Aristotle (as was natural in his line of discussion) turns his attention entirely to the *evidence* of the inference; and overlooks a step which is of far more importance to our knowledge, namely, the *invention* of the second extreme term. In the above instance, the particular luminaries, Mercury, Venus, Mars, are one logical *Extreme;* the general designation Planets is the *Middle Term;* but having these before us, how do we come to think of *description of ellipses,* which is the other Extreme of the syllogism? When we have once invented this 'second Extreme Term,' we may, or may not, be satisfied with the evidence of the syllogism; we may, or may not, be convinced that, so far as this property goes, the extremes are co-extensive with the middle term;[5] but the *statement* of the syllogism is the important step in science. We know how long Kepler laboured, how hard he fought, how many devices he tried, before he hit upon this *Term,* the Elliptical Motion. He rejected, as we know, many other 'second extreme Terms,' for example, various combinations of epicyclical constructions, because they did not represent with sufficient accuracy the special facts of observation. When he had established his premiss, that 'Mars does describe an Ellipse about the Sun,' he does not hesitate to *guess* at least that, in this respect, he might *convert* the other premiss, and assert that 'All the Planets do what Mars does.' But the main business was, the inventing and verifying the proposition respecting the Ellipse. The Invention of the Conception was the great step in the *discovery;* the Verification of the Proposition was the great step in the *proof* of the discovery. If

Logic consists in pointing out the conditions of proof, the Logic of Induction must consist in showing what are the conditions of proof, in such inferences as this: but this subject must be pursued in the next chapter; I now speak principally of the act of *Invention,* which is requisite in every inductive inference.

4. Although in every inductive inference, an act of invention is requisite, the act soon slips out of notice. Although we bind together facts by superinducing upon them a new Conception, this Conception, once introduced and applied, is looked upon as inseparably connected with the facts, and necessarily implied in them. Having once had the phenomena bound together in their minds in virtue of the Conception, men can no longer easily restore them back to the detached and incoherent condition in which they were before they were thus combined. The pearls once strung, they seem to form a chain by their nature. Induction has given them a unity which it is so far from costing us an effort to preserve, that it requires an effort to imagine it dissolved. For instance, we usually represent to ourselves the Earth as *round,* the Earth and the Planets as *revolving* about the Sun, and as *drawn* to the Sun by a Central Force; we can hardly understand how it could cost the Greeks, and Copernicus, and Newton, so much pains and trouble to arrive at a view which to us is so familiar. These are no longer to us Conceptions caught hold of and kept hold of by a severe struggle; they are the simplest modes of conceiving the facts: they are really Facts. We are willing to *own* our obligation to those discoverers, but we hardly *feel* it: for in what other manner (we ask in our thoughts) could we represent the facts to ourselves?

Thus we see why it is that this step of which we now speak, the Invention of a new Conception in every inductive inference, is so generally overlooked that it has hardly been noticed by preceding philosophers. When once performed by the discoverer, it takes a fixed and permanent place in the understanding of every one. It is a thought which, once breathed forth, permeates all men's minds. All fancy they nearly or quite knew it before. It oft was thought, or almost thought, though never till now expressed. Men accept it and retain it, and know it cannot be taken from them, and look upon it as their own. They will not and cannot part with it, even though they may deem it trivial and obvious. It is a secret, which once uttered, cannot be recalled, even though it be despised by those to whom it is imparted. As soon as the leading term of a new theory has been pronounced and understood, all the phenomena change their aspect. There is a standard to which we cannot help referring them. We cannot fall back into the helpless and bewildered state in which we gazed at them when we possessed no principle which gave them unity. Eclipses arrive in mysterious confusion: the notion of a *Cycle* dispels the mystery. The Planets perform a tangled and mazy dance; but *Epicycles* reduce the maze to order. The Epicycles themselves run into confusion; the conception of an *Ellipse* makes all clear and simple. And thus from stage to stage, new elements of intelligible order are introduced. But this intelligible order is so completely adopted by the human understanding, as to seem part of its texture. Men ask Whether Eclipses follow a Cycle; Whether the Planets describe Ellipses; and they imagine that so long as they do not *answer* such questions rashly, they take nothing for granted. They do not recollect how much they assume in *asking* the question:—how far the conceptions of Cycles and of Ellipses are beyond the visible surface of the celestial phenomena:—how many ages elapsed, how much thought, how much observation, were needed, before men's thoughts were fashioned into the words which they now so familiarly use. And thus they treat the subject, as we have seen Aristotle treating it; as if it were a question, not of invention, but of proof; not of substance, but of form: as if the main thing were not *what* we assert, but *how* we assert it. But for our purpose, it is requisite to bear in mind the feature we have thus attempted to mark; and to recollect that, in every inference by induction, there is a Conception supplied by the mind and superinduced upon the Facts.

5. In collecting scientific truths by Induction, we often find (as has already been observed) a Definition and a Proposition established at the same time,—introduced together, and mutually dependent on each other. The combination of the two constitutes the Inductive act; and we may consider the Definition as representing the superinduced Conception, and the Proposition as exhibiting the Colligation of Facts.

### *Sect. II. Use of Hypotheses*

6. To discover a Conception of the mind which will justly represent a train of observed facts is, in some measure, a process of con-

jecture, as I have stated already; and as I then observed, the business of conjecture is commonly conducted by calling up before our minds several suppositions, and selecting that one which most agrees with what we know of the observed facts. Hence he who has to discover the laws of nature may have to invent many suppositions before he hits upon the right one; and among the endowments which lead to his success, we must reckon that fertility of invention which ministers to him such imaginary schemes, till at last he finds the one which conforms to the true order of nature. A facility in devising hypotheses, therefore, is so far from being a fault in the intellectual character of a discoverer, that it is, in truth, a faculty indispensable to his task. It is, for his purpose, much better that he should be too ready in contriving, too eager in pursuing systems which promise to introduce law and order among a mass of unarranged facts, than that he should be barren of such inventions and hopeless of such success. Accordingly, as we have already noticed, great discoverers have often invented hypotheses which would not answer to all the facts, as well as those which would; and have fancied themselves to have discovered laws, which a more careful examination of the facts overturned.

The tendencies of our speculative nature,[6] carrying us onwards in pursuit of symmetry and rule, and thus producing all true theories, perpetually show their vigour by overshooting the mark. They obtain something, by aiming at much more. They detect the order and connexion which exist, by conceiving imaginary relations of order and connexion which have no existence. Real discoveries are thus mixed with baseless assumptions; profound sagacity is combined with fanciful conjucture; not rarely, or in peculiar instances, but commonly, and in most cases; probably in all, if we could read the thoughts of discoverers as we read the books of Kepler. To try wrong guesses is, with most persons, the only way to hit upon right ones. The character of the true philosopher is, not that he never conjectures hazardously, but that his conjectures are clearly conceived, and brought into rigid contact with facts. He sees and compares distinctly the Ideas and the Things;—the relations of his notions to each other and to phenomena. Under these conditions, it is not only excusable, but necessary for him, to snatch at every semblance of general rule,—to try all promising forms of simplicity and symmetry.

Hence advances in knowledge[7] are not commonly made without the previous exercise of some boldness and license in guessing. The discovery of new truths requires, undoubtedly, minds careful and scrupulous in examing what is suggested; but it requires, no less, such as are quick and fertile in suggesting. What is Invention, except the talent of rapidly calling before us the many possibilities, and selecting the appropriate one? It is true, that when we have rejected all the inadmissible suppositions, they are often quickly forgotten; and few think it necessary to dwell on these discarded hypotheses, and on the process by which they were condemned. But all who discover truths, must have reasoned upon many errours to obtain each truth; every accepted doctrine must have been one chosen out of many candidates. If many of the guesses of philosophers of bygone times now appear fanciful and absurd, because time and observation have refuted them, others, which were at the time equally gratuitous, have been confirmed in a manner which makes them appear marvellously sagacious. To form hypotheses, and then to employ much labour and skill in refuting them, if they do not succeed in establishing them, is a part of the usual process of inventive minds. Such a proceeding belongs to the *rule* of the genius of discovery, rather than (as has often been taught in modern times) to the *exception*.

7. But if it be an advantage for the discoverer of truth that he be ingenious and fertile in inventing hypotheses which may connect the phenomena of nature, it is indispensably requisite that he be diligent and careful in comparing his hypotheses with the facts, and ready to abandon his invention as soon as it appears that it does not agree with the course of actual occurrences. This constant comparison of his own conceptions and supposition with observed facts under all aspects, forms the leading employment of the discoverer: this candid and simple love of truth, which makes him willing to suppress the most favourite production of his own ingenuity as soon as it appears to be at variance with realities, constitutes the first characteristic of his temper. He must have neither the blindness which cannot, nor the obstinacy which will not, perceive the discrepancy of his fancies and his facts. He must allow no indolence, or partial views, or self-complacency, or delight in seeming demonstration, to make him tenacious of the schemes which he devises, any further than they are confirmed by

their accordance with nature. The framing of hypotheses is, for the inquirer after truth, not the end, but the beginning of his work. Each of his systems is invented, not that he may admire it and follow it into all its consistent consequences, but that he may make it the occasion of a course of active experiment and observation. And if the results of this process contradict his fundamental assumptions, however ingenious, however symmetrical, however elegant his system may be, he rejects it without hesitation. He allows no natural yearning for the offspring of his own mind to draw him aside from the higher duty of loyalty to his sovereign, Truth: to her he not only gives his affections and his wishes, but strenuous labour and scrupulous minuteness of attention.

We may refer to what we have said of Kepler, Newton, and other eminent philosophers, for illustrations of this character. In Kepler we have remarked[8] the courage and perseverance with which he undertook and executed the task of computing his own hypotheses: and, as a still more admirable characteristic, that he never allowed the labour he had spent upon any conjecture to produce any reluctance in abandoning the hypothesis, as soon as he had evidence of its inaccuracy. And in the history of Newton's discovery that the moon is retained in her orbit by the force of gravity, we have noticed the same moderation in maintaining the hypothesis, after it had once occurred to the author's mind. The hypothesis required that the moon should fall from the tangent of her orbit every second through a space of sixteen feet; but according to his first calculations it appeared that in fact she only fell through a space of thirteen feet in that time. The difference seems small, the approximation encouraging, the theory plausible; a man in love with his own fancies would readily have discovered or invented some probable cause of the difference. But Newton acquiesced in it as a disproof of his conjecture, and 'laid aside at that time any further thoughts of this matter.'[9]

8. It has often happened that those who have undertaken to instruct mankind have not possessed this pure love of truth and comparative indifference to the maintenance of their own inventions. Men have frequently adhered with great tenacity and vehemence to the hypotheses which they have once framed; and in their affection for these, have been prone to overlook, to distort, and to misinterpret

facts. In this manner, *Hypotheses* have so often been prejudicial to the genuine pursuit of truth, that they have fallen into a kind of obloquy; and have been considered as dangerous temptations and fallacious guides. Many warnings have been uttered against the fabrication of hypotheses, by those who profess to teach philosophy; many disclaimers of such a course by those who cultivate science.

Thus we shall find Bacon frequently discommending this habit, under the name of 'anticipation of the mind,' and Newton thinks it necessary to say emphatically 'hypotheses non fingo.' It has been constantly urged that the inductions by which sciences are formed must be *cautious* and *rigorous;* and the various imaginations which passed through Kepler's brain, and to which he has given utterance have been blamed or pitied, as lamentable instances of an unphilosophical frame of mind. Yet it has appeared in the preceding remarks that hypotheses rightly used are among the helps, far more than the dangers, of science;—that scientific induction is not a 'cautious' or a 'rigorous' process in the sense of *abstaining from* such suppositions, but in *not adhering* to them till they are confirmed by fact, and in carefully seeking from facts confirmation or refutation. Kepler's distinctive character was, not that he was peculiarly given to the construction of hypotheses, but that he narrated with extraordinary copiousness and candour the course of his thoughts, his labours, and his feelings. In the minds of most persons, as we have said, the inadmissible suppositions, when rejected, are soon forgotten: and thus the trace of them vanishes from the thoughts, and the successful hypothesis alone holds its place in our memory. But in reality, many other transient suppositions must have been made by all discoverers;—hypotheses which are not afterwards asserted as true systems, but entertained for an instant;—'tentative hypotheses,' as they have been called. Each of these hypotheses is followed by its corresponding train of observations, from which it derives its power of leading to truth. The hypothesis is like the captain, and the observations like the soldiers of an army: while he appears to command them, and in this way to work his own will, he does in fact derive all his power of conquest from their obedience, and becomes helpless and useless if they mutiny.

Since the discoverer has thus constantly to work his way onwards by means of hypotheses, false and true, it is highly important for him to possess talents and means for rapidly *testing* each supposition as it offers itself. In this as in other parts of the work of discovery, success has in general been mainly owing to the native ingenuity and sagacity of the discoverer's mind. Yet some Rules tending to further this object have been delivered by eminent philosophers, and some others may perhaps be suggested. Of these we shall here notice only some of the most general, leaving for a future chapter the consideration of some more limited and detailed processes by which, in certain cases, the discovery of the laws of nature may be materially assisted.

#### *Sect. III. Tests of Hypotheses*

9. A maxim which it may be useful to recollect is this,—that *hypotheses may often be of service to science, when they involve a certain portion of incompleteness, and even of errour.* The object of such inventions is to bind together facts which without them are loose and detached; and if they do this, they may lead the way to a perception of the true rule by which the phenomena are associated together, even if they themselves somewhat misstate the matter. The imagined arrangement enables us to contemplate, as a whole, a collection of special cases which perplex and overload our minds when they are considered in succession; and if our scheme has so much of truth in it as to conjoin what is really connected, we may afterwards duly correct or limit the mechanism of this connexion. If our hypothesis renders a reason for the agreement of cases really similar, we may afterwards find this reason to be false, but we shall be able to translate it into the language of truth.

A conspicuous example of such an hypothesis,—one which was of the highest value to science, though very incomplete, and as a representation of nature altogether false,—is seen in the *Doctrine of epicycles* by which the ancient astronomers explained the motions of the sun, moon, and planets. This doctrine connected the places and velocities of these bodies at particular times in a manner which was, in its general features, agreeable to nature. Yet this doctrine was erroneous in its assertion of the *circular* nature of all the celestial motions, and in making the heavenly bodies revolve *round the earth.* It was, however, of immense value to the progress of astronomical science; for it enabled men to express and reason upon many important truths which they discovered respecting the motion of the stars, up to the time of Kepler. Indeed we can hardly imagine that astronomy could, in its outset, have made so great a progress under any other form, as it did in consequence of being cultivated in this shape of the incomplete and false *epicyclical hypothesis*.

We may notice another instance of an exploded hypothesis, which is generally mentioned only to be ridiculed, and which undoubtedly is both false in the extent of its assertion, and unphilosophical in its expression; but which still, in its day, was not without merit. I mean the doctrine of *Nature's horrour of a vacuum* (*fuga vacui*), by which the action of siphons and pumps and many other phenomena were explained, till Mersenne and Pascal taught a truer doctrine. This hypothesis was of real service; for it brought together many facts which really belong to the same class, although they are very different in their first aspect. A scientific writer of modern times[10] appears to wonder that men did not at once divine the weight of the air, from which the phenomena formerly ascribed to the *fuga vacui* really result. 'Loaded, compressed by the atmosphere,' he says, 'they did not recognize its action. In vain all nature testified that air was elastic and heavy; they shut their eyes to her testimony. The water rose in pumps and flowed in siphons at that time, as it does at this day. They could not separate the boards of a pair of bellows of which the holes were stopped; and they could not bring together the same boards without difficulty, if they were at first separated. Infants sucked the milk of their mothers; air entered rapidly into the lungs of animals at every inspiration; cupping-glasses produced tumours on the skin; and in spite of all these striking proofs of the weight and elasticity of the air, the ancient philosophers maintained resolutely that air was light, and explained all these phenomena by the horrour which they said nature had for a vacuum.' It is curious that it should not have occurred to the author while writing this, that if these facts, so numerous and various, can all be accounted for by *one* principle, there is a strong presumption that the principle is not altogether baseless. And in reality is it not true that nature *does* abhor a vacuum, and does all she can to avoid it? No doubt this power is not unlimited; and moreover we can trace it to a mechanical cause, the pressure of the circumambient air. But the tendency, arising from this pressure, which the bodies surrounding a space void of air have to rush into it, may be expressed, in no extravagant or unintelligible manner, by saying that nature has a repugnance to a vacuum.

That imperfect and false hypotheses, though they may thus explain *some* phenomena, and may be useful in the progress of science, cannot explain *all* phenomena;—and that we are never to rest in our labours or acquiesce in our results, till we have found some view of the subject which *is* consistent with *all* the observed facts;—will of course be understood. We shall afterwards have to speak of the other steps of such a progress.

10. Thus the hypotheses which we accept ought to explain phenomena which we have observed. But they ought to do more than this: our hypotheses ought to *foretel* phenomena which have not yet been observed; at least all phenomena of the same kind as those which the hypothesis was invented to explain. For our assent to the hypothesis implies that it is held to be true of all particular instances. That these cases belong to past or to future times, that they have or have not already occurred, makes no difference in the applicability of the rule to them. Because the rule prevails, it includes all cases; and will determine them all, if we can only calculate its real consequences. Hence it will predict the results of new combinations, as well as explain the appearances which have occurred in old ones. And that it does this with certainty and correctness, is one mode in which the hypothesis is to be verified as right and useful.

The scientific doctrines which have at various periods been established have been verified in this manner. For example, the *Epicyclical Theory* of the heavens was confirmed by its *predicting* truly eclipses of the sun and moon, configurations of the planets, and other celestial phenomena; and by its leading to the construction of Tables by which the places of the heavenly bodies were given at every moment of time. The truth and accuracy of these predictions were a proof that the hypothesis was valuable, and, at least to a great extent, true; although, as was afterwards found, it involved a false representation of the structure of the heavens. In like manner, the discovery of the *Laws of Refraction* enabled mathematicians to *predict*, by calculation, what would be the effect of any new form or combination of transparent lenses. Newton's hypothesis of *Fits of Easy Transmission and*

*Easy Reflection* in the particles of light, although not confirmed by other kinds of facts, involved a true statement of the law of the phenomena which it was framed to include, and served to *predict* the forms and colours of thin plates for a wide range of given cases. The hypothesis that Light operates by *Undulations* and *Interferences*, afforded the means of *predicting* results under a still larger extent of conditions. In like manner in the progress of chemical knowledge, the doctrine of *Phlogiston* supplied the means of *foreseeing* the consequence of many combinations of elements, even before they were tried; but the *Oxygen Theory*, besides affording predictions, at least equally exact, with regard to the general results of chemical operations, included all the facts concerning the relations of weight of the elements and their compounds, and enabled chemists to *foresee* such facts in untried cases. And the Theory of *Electromagnetic Forces*, as soon as it was rightly understood, enabled those who had mastered it to *predict* motions such as had not been before observed, which were accordingly found to take place.

Men cannot help believing that the laws laid down by discoverers must be in a great measure identical with the real laws of nature, when the discoverers thus determine effects beforehand in the same manner in which nature herself determines them when the occasion occurs. Those who can do this, must, to a considerable extent, have detected nature's secret;—must have fixed upon the conditions to which she attends, and must have seized the rules by which she applies them. Such a coincidence of untried facts with speculative assertions cannot be the work of chance, but implies some large portion of truth in the principles on which the reasoning is founded. To trace order and law in that which has been observed, may be considered as interpreting what nature has written down for us, and will commonly prove that we understand her alphabet. But to predict what has not been observed, is to attempt ourselves to use the legislative phrases of nature; and when she responds plainly and precisely to that which we thus utter, we cannot but suppose that we have in a great measure made ourselves masters of the meaning and structure of her language. The prediction of results, even of the same kind as those which have been observed, in new cases, is a proof of real success in our inductive processes.

11. We have here spoken of the prediction of facts *of the same kind* as those from which our rule was collected. But the evidence in favour of our induction is of a much higher and more forcible character when it enables us to explain and determine cases of a *kind different* from those which were contemplated in the formation of our hypothesis. The instances in which this has occurred, indeed, impress us with a conviction that the truth of our hypothesis is certain. No accident could give rise to such an extraordinary coincidence. No false supposition could, after being adjusted to one class of phenomena, exactly represent a different class, where the agreement was unforeseen and uncontemplated. That rules springing from remote and unconnected quarters should thus leap to the same point, can only arise from *that* being the point where truth resides.

Accordingly the cases in which inductions from classes of facts altogether different have thus *jumped together*, belong only to the best established theories which the history of science contains. And as I shall have occasion to refer to this peculiar feature in their evidence, I will take the liberty of describing it by a particular phrase; and will term it the *Consilience of Inductions*.

It is exemplified principally in some of the greatest discoveries. Thus it was found by Newton that the doctrine of the Attraction of the Sun varying according to the Inverse Square of this distance, which explained Kepler's *Third Law*, of the proportionality of the cubes of the distance to the squares of the periodic times of the planets, explained also his *First* and *Second Laws*, of the elliptical motion of each planet; although no connexion of these laws had been visible before. Again, it appeared that the force of Universal Gravitation, which had been inferred from the *Perturbations* of the moon and planets by the sun and by each other, also accounted for the fact, apparently altogether dissimilar and remote, of the *Precession of the equinoxes*. Here was a most striking and surprising coincidence, which gave to the theory a stamp of truth beyond the power of ingenuity to counterfeit. In like manner in Optics; the hypothesis of alternate Fits of easy Transmission and Reflection would explain the colours of thin plates, and indeed was devised and adjusted for that very purpose; but it could give no account of the phenomena of the fringes of shadows. But the doctrine of Interferences, constructed at first with reference to phenomena of the nature of the *Fringes,* explained also the *Colours of thin plates* better than the supposition of the Fits invented for that very purpose. And we have in Physical Optics another example of the same kind, which is quite as striking as the explanation of Precession by inferences from the facts of Perturbation. The doctrine of Undulations propagated in a Spheroidal Form was contrived at first by Huyghens, with a view to explain the laws of *Double Refraction* in calc-spar; and was pursued with the same view by Fresnel. But in the course of the investigation it appeared, in a most unexpected and wonderful manner, that this same doctrine of spheroidal undulations, when it was so modified as to account for the directions of the two refracted rays, accounted also for the positions of their *Planes of Polarization,*[11] a phenomenon which, taken by itself, it had perplexed previous mathematicians, even to represent.

The Theory of Universal Gravitation, and of the Undulatory Theory of Light, are, indeed, full of examples of this Consilience of Inductions. With regard to the latter, it has been justly asserted by Herschel, that the history of the undulatory theory was a succession of *felicities.*[12] And it is precisely the unexpected coincidences of results drawn from distant parts of the subject which are properly thus described. Thus the Laws of the *Modification of polarization* to which Fresnel was led by his general views, accounted for the Rule respecting the *Angle at which light is polarized,* discovered by Sir D. Brewster.[13] The conceptions of the theory pointed out peculiar *Modifications* of the phenomena when *Newton's rings* were produced by polarised light, which modifications were ascertained to take place in fact, by Arago and Airy.[14] When the beautiful phenomena of *Dipolarized light* were discovered by Arago and Biot, Young was able to declare that they were reducible to the general laws of Interference which he had already established.[15] And what was no less striking a confirmation of the truth of the theory, *Measures* of the same element deduced from various classes of facts were found to coincide. Thus the Length of a luminiferous undulation, calculated by Young from the measurement of *Fringes* of shadows, was found to agree very nearly with the previous calculation from the colours of *Thin plates.*[16]

No example can be pointed out, in the whole history of science, so far as I am aware, in which this Consilience of Inductions has given testimony in favour of an hypothesis afterwards discovered to be false. If we take one class of facts only, knowing the law which they follow, we may construct an hypothesis, or perhaps several, which may represent them: and as new circumstances are discovered, we may often adjust the hypothesis so as to correspond to these also. But when the hypothesis, of itself and without adjustment for the purpose, gives us the rule and reason of a class of facts not contemplated in its construction, we have a criterion of its reality, which has never yet been produced in favour of falsehood.

12. In the preceding Article I have spoken of the hypothesis with which we compare our facts as being framed *all at once,* each of its parts being included in the original scheme. In reality, however, it often happens that the various suppositions which our system contains are *added* upon occasion of different researches. Thus in the Ptolemaic doctrine of the heavens, new epicycles and eccentrics were added as new inqualities of the motions of the heavenly bodies were discovered; and in the Newtonian doctrine of material rays of light, the supposition that these rays had 'fits,' was added to explain the colours of thin plates; and the supposition that they had 'sides' was introduced on occasion of the phenomena of polarization. In like manner other theories have been built up of parts devised at different times.

This being the mode in which theories are often framed, we have to notice a distinction which is found to prevail in the progress of true and false theories. In the former class all the additional suppositions *tend to simplicity* and harmony; the new suppositions resolve themselves into the old ones, or at least require only some easy modification of the hypothesis first assumed: the system becomes more coherent as it is further extended. The elements which we require for explaining a new class of facts are already contained in our system. Different members of the theory run together, and we have thus a constant convergence to unity. In false theories, the contrary is the case. The new suppositions are something altogether additional;—not suggested by the original scheme; perhaps difficult to reconcile with it. Every such addition adds to the complexity of the hypothetical system, which at last becomes unmanageable, and is compelled to surrender its place to some simpler explanation.

Such a false theory for example was the ancient doctrine of eccentrics and epicycles. It explained the general succession of the Places of the Sun, Moon, and Planets; it would not have explained the proportion of their Magnitudes at different times, if these could have been accurately observed; but this the ancient astronomers were unable to do. When, however, Tycho and other astronomers came to be able to observe the planets accurately in all positions, it was found that *no* combination of *equable* circular motions would exactly represent all the observations. We may see, in Kepler's works, the many new modifications of the epicyclical hypothesis which offered themselves to him; some of which would have agreed with the phenomena with a certain degree of accuracy, but not with so great a degree as Kepler, fortunately for the progress of science, insisted upon obtaining. After these epicycles had been thus accumulated, they all disappeared and gave way to the simpler conception of an *elliptical* motion. In like manner, the discovery of new inequalities in the Moon's motions encumbered her system more and more with new machinery, which was at last rejected all at once in favour of the *elliptical* theory. Astronomers could not but suppose themselves in a wrong path, when the prospect grew darker and more entangled at every step.

Again; the Cartesian system of Vortices might be said to explain the primary phenomena of the revolutions of planets about the sun, and satellites about the planets. But the elliptical form of the orbits required new suppositions. Bernoulli ascribed this curve to the shape of the planet, operating on the stream of the vortex in a manner similar to the rudder of a boat. But then the motions of the aphelia, and of the nodes,—perturbations,—even the action of gravity towards the earth,—could not be accounted for without new and independent suppositions. Here was none of the simplicity of truth. The theory of Gravitation, on the other hand, became more simple as the facts to be explained became more numerous. The attraction of the sun accounted for the motions of the planets; the attraction of the planets was the cause of the motion of the satellites. But this being assumed, the perturbations, and the motions of the nodes and aphelia, only made it requisite to extend the attraction of the sun to the satellites, and that of the planets to each other:—the tides, the speroidal form of the earth, the precession, still required nothing more than that the moon and sun should attract the parts of the earth, and that these should attract each other;—so that all the suppositions resolved themselves into the single one, of the universal gravitation of all matter. It is difficult to imagine a more convincing manifestation of simplicity and unity.

Again, to take an example from another science;—the doctrine of Phlogiston brought together many facts in a very plausible manner,—combustion, acidification, and others,—and very naturally prevailed for a while. But the balance came to be used in chemical operations, and the facts of weight as well as of combination were to be accounted for. On the phlogistic theory, it appeared that this could not be done without a new supposition, and *that,* a very strange one;—that phlogiston was an element not only not heavy, but absolutely light, so that it diminished the weight of the compounds into which it entered. Some chemists for a time adopted this extravagent view; but the wiser of them saw, in the necessity of such a supposition to the defence of the theory, an evidence that the hypothesis of an element *phlogiston* was erroneous. And the opposite hypothesis, which taught that oxygen was subtracted, and not phlogiston added, was accepted because it required no such novel and inadmissible assumption.

Again, we find the same evidence of truth in the progress of the Undulatory Theory of light, in the course of its application from one class of facts to another. Thus we explain Reflection and Refraction by undulations; when we come to Thin Plates, the requisite 'fits' are already involved in our fundamental hypothesis, for they are the length of an undulation: the phenomena of Diffraction also require such intervals; and the intervals thus required agree exactly with the others in magnitude, so that no new property is needed. Polarization for a moment appears to require some new hypothesis; yet this is hardly the case; for the direction of our vibrations is hitherto arbitrary:—we allow polarization to decide it, and we suppose the undulations to be transverse. Having done this for the sake of Polarization, we turn to the phenomena of Double Refraction, and inquire what new hypothesis they require. But the answer is, that they require none: the supposition of transverse vibrations, which we have made in order to explain Polarization, gives us also the law of Double Refraction. Truth may give rise to such a coincidence; falsehood cannot. Again, the facts of Dipolarization come into view. But they hardly require any new assumption; for the difference of optical elasticity of crystals in different directions, which is already assumed in uniaxal crystals,[17] is extended to biaxal exactly according to the law of symmetry; and this being done, the laws of the phenomena, curious and complex as they are, are fully explained. The phenomena of Circular Polarization by internal reflection, instead of requiring a new hypothesis, are found to be given by an interpretation of an apparently inexplicable result of an old hypothesis. The Circular Polarization of Quartz and its Double Refraction does indeed appear to require a new assumption, but still not one which at all disturbs the form of the theory; and in short, the whole history of this theory is a progress, constant and steady, often striking and startling, from one degree of evidence and consistence to another of higher order.

In the Emission Theory, on the other hand, as in the theory of solid epicycles, we see what we may consider as the natural course of things in the career of a false theory. Such a theory may, to a certain extent, explain the phenomena which it was at first contrived to meet; but every new class of facts requires a new supposition—an addition to the machinery: and as observation goes on, these incoherent appendages accumulate, till they overwhelm and upset the original frame-work. Such has been the hypothesis of the Material Emission of light. In its original form, it explained Reflection and Refraction: but the colours of Thin Plates added to it the Fits of easy Transmission and Reflection; the phenomena of Diffraction further invested the emitted particles with complex laws of Attraction and Repulsion; Polarization gave them Sides: Double Refraction subjected them to peculiar Forces emanating from the axes of the crystal: finally, Dipolarization loaded them with the complex and unconnected contrivance of Moveable Polarization: and even when all this had been done, additional mechanism was wanting. There is here no unexpected success, no happy coincidence, no convergence of principles from remote quarters. The philosopher builds the machine, but its parts do not fit. They hold together only while he presses them. This is not the character of truth.

As another example of the application of the Maxim now under consideration, I may perhaps be allowed to refer to the judgment which, in the History of Thermotics, I have ventured to give respecting Laplace's Theory of Gases. I have stated,[18] that we cannot help forming an unfavourable judgment of this theory, by looking for that great characteristic of true theory; namely, that the hypotheses which were assumed to account for *one class* of facts are found to explain *another class* of a different nature. Thus Laplace's first suppositions explain the connexion of Compression with Density, (the law of Boyle and Mariotte,) and the connexion of Elasticity with Heat, (the law of Dalton and Gay Lussac). But the theory requires other assumptions when we come to Latent Heat; and yet these new assumptions produce no effect upon the calculations in any application of the theory. When the hypothesis, constructed with reference to the Elasticity and Temperature, is applied to another class of facts, those of Latent Heat, we have no Simplication of the Hypothesis, and therefore no evidence of the truth of the theory.

13. The last two sections of this chapter direct our attention to two circumstances, which tend to prove, in a manner which we may term irresistible, the truth of the theories which they characterize:—the *Consilience of Inductions* from different and separate classes of facts;—and the progressive *Simplification of the Theory* as it is extended to new cases. These two Characters are, in fact, hardly different; they are exemplified by the same cases. For if these Inductions, collected from one class of facts, supply an unexpected explanation of a new class, which is the case first spoken of, there will be no need for new machinery in the hypothesis to apply it to the newly-contemplated facts; and thus, we have a case in which the system does not become more complex when its application is extended to a wider field, which was the character of true theory in its second aspect. The Consiliences of our Inductions give rise to a constant Convergence of our Theory towards Simplicity and Unity.

But, moreover, both these cases of the extension of the theory, without difficulty or new suppositions, to a wider range and to new classes of phenomena, may be conveniently considered in yet another point of view; namely, as successive steps by which we gradually ascend in our speculative views to a higher and higher point of generality. For when the theory, either by the concurrence of two indications, or by an extension without complication, has included a new range of phenomena, we have, in fact, a new induction of a more general kind, to which the inductions formerly obtained are subordinate, as particular cases to a general proposition. We have in such examples, in short, an instance of *successive generalization.* This is a subject of great importance, and deserving of being well illustrated; it will come under our notice in the next chapter.

## CHAPTER VI
## OF THE LOGIC OF INDUCTION

***Aphorism XVII***

*The* Logic of Induction *consists in stating the Facts and the Inference in such a manner, that the Evidence of the Inference is manifest; just as the Logic of Deduction consists in stating the Premises and the Conclusion in such a manner that the Evidence of the Conclusion is manifest.*

***Aphorism XVIII***

*The Logic of Deduction is exhibited by means of a certain Formula; namely, a Syllogism; and every train of deductive reasoning, to be demonstrative, must be capable of resolution into a series of such Formulae legitimately constructed. In like manner, the Logic of Induction may be exhibited by means of certain* Formulae; *and every train of inductive inference, to be sound, must be capable of resolution into a scheme of such Formulae, legitimately constructed.*

***Aphorism XIX***

*The* inductive act of thought *by which several Facts are colligated into one Proposition, may be expressed by saying:* The several Facts are exactly expressed as one Fact, if, and only if, we adopt the Conceptions and the Assertion *of the Proposition.*

***Aphorism XX***

*The One Fact, thus inductively obtained from several Facts, may be combined with other Facts, and colligated with them by a new act of Induction. This process may be indefinitely repeated: and these successive processes are the* Steps *of Induction, or of* generalization, *from the lowest to the highest.*

*Aphorism XXI*

*The relation of the successive Steps of Induction may be exhibited by means of an* Inductive Table, *in which the several Facts are indicated, and tied together by a Bracket, and the Inductive Inference placed on the other side of the Bracket; and this arrangement repeated, so as to form a genealogical Table of each Induction, from the lowest to the highest.*

*Aphorism XXII*

*The Logic of Induction is the* Criterion of Truth *inferred from Facts, as the Logic of Deduction is the Criterion of Truth deduced from necessary Principles. The Inductive Table enables us to apply such a Criterion; for we can determine whether each Induction is verified and justified by the Facts which its Bracket includes; and if each induction in particular be sound, the highest, which merely combines them all, must necessarily be sound also.*

*Aphorism XXIII*

*The distinction of* Fact *and* Theory *is only relative. Events and phenomena, considered as Particulars which may be colligated by Induction, are* Facts; *considered as Generalities already obtained by colligation of other Facts, they are* Theories. *The same event or phenomenon is a Fact or a Theory, according as it is considered as standing on one side or the other of the Inductive Bracket.*

1. The subject to which the present chapter refers is described by phrases which are at the present day familiarly used in speaking of the progress of knowledge. We hear very frequent mention of *ascending from particular to general* propositions, and from these to propositions still more general;—of truths *included* in other truths of a higher degree of generality;—of different *stages of generalization;*—and of the *highest step* of the process of discovery, to which all others are subordinate and preparatory. As these expressions, so familiar to our ears, especially since the time of Francis Bacon, denote, very significantly, processes and relations which are of great importance in the formation of science, it is necessary for us to give a clear account of them, illustrated with general exemplifications; and this we shall endeavour to do.

We have, indeed, already explained that science consists of Propositions which include the Facts from which they were collected; and other wider Propositions, collected in like manner from the former, and including them. Thus, that the stars, the moon, the sun, rise, culminate, and set, are facts *included* in the proposition that the heavens, carrying with them all the celestial bodies, have a diurnal revolution about the axis of the earth. Again, the observed monthly motions of the moon, and the annual motions of the sun, are *included* in certain propositions concerning the movements of those luminaries with respect to the stars. But all these propositions are really included in the doctrine that the earth, revolving on its axis, moves round the sun, and the moon round the earth. These movements, again, considered as facts, are explained and *included* in the statement of the forces which the earth exerts upon the moon, and the sun upon the earth. Again, this doctrine of the forces of these three bodies is *included* in the assertion, that all the bodies of the solar system, and all parts of matter, exert forces, each upon each. And we might easily show that all the leading facts in astronomy are comprehended in the same generalization. In like manner with regard to any other science, so far as its truths have been well established and fully developed, we might show that it consists of a gradation of propositions, proceeding from the most special facts to the most general theoretical assertions. We shall exhibit this gradation in some of the principal branches of science.

2. This gradation of truths, successively included in other truths, may be conveniently represented by *Tables* resembling the genealogical tables by which the derivation of descendants from a common ancestor is exhibited; except that it is proper in this case to invert the form of the Table, and to make it converge to unity downwards instead of upwards, since it has for its purpose to express, not the derivation of many from one, but the collection of one truth from many things. Two or more co-ordinate facts or propositions may be ranged side by side, and joined by some mark of connexion, (a bracket, as ⏟ or ⎣⎦ ,) beneath which may be placed the more general proposition which is collected by induction from the former. Again, propositions co-ordinate with this more general one may be placed on a level with it; and the combination of these, and the result of the

combination, may be indicated by brackets in the same manner; and so on, through any number of gradations. By this means the streams of knowledge from various classes of facts will constantly run together into a smaller and smaller number of channels; like the confluent rivulets of a great river, coming together from many sources, uniting their ramifications so as to form larger branches, these again uniting in a single trunk. The *genealogical tree* of each great portion of science, thus formed, will contain all the leading truths of the science arranged in their due co-ordination and subordination. Such Tables, constructed for the sciences of Astronomy and of Optics, will be given at the end of this chapter. [The Tables appear following page 180.]

3. The union of co-ordinate propositions into a proposition of a higher order, which occurs in this Tree of Science wherever two twigs unite in one branch, is, in each case, an example of *Induction.* The single proposition is collected by the process of induction from its several members. But here we may observe, that the image of a mere *union* of the parts at each of these points, which the figure of a tree or a river presents, is very inadequate to convey the true state of the case; for in Induction, as we have seen, besides mere collection of particulars, there is always a *new conception*, a principle of connexion and unity, supplied by the mind, and superinduced upon the particulars. There is not merely a juxta-position of materials, by which the new proposition contains all that its component parts contained; but also a formative act exerted by the understanding, so that these materials are contained in a new shape. We must remember, therefore, that our Inductive Tables, although they represent the elements and the order of these inductive steps, do not fully represent the whole signification of the process in each case.

4. The principal features of the progress of science spoken of in the last chapter are clearly exhibited in these Tables; namely, the *Consilience of Inductions,* and the constant Tendency to Simplicity observable in true theories. Indeed in all cases in which, from propositions of considerable generality, propositions of a still higher degree are obtained, there is a convergence of inductions; and if in one of the lines which thus converge, the steps be rapidly and suddenly made in order to meet the other line, we may consider that we have an ex-

ample of Consilience. Thus when Newton had collected, from Kepler's Laws, the Central Force of the sun, and from these, combined with other facts, the Universal Force of all the heavenly bodies, he suddenly turned round to include in his generalization the Precession of the Equinoxes, which he declared to arise from the attraction of the sun and moon upon the protuberant part of the terrestrial spheroid. The apparent remoteness of this fact, in its nature, from the other facts with which he thus associated it, causes this part of his reasoning to strike us as a remarkable example of *Consilience*. Accordingly, in the Table of Astronomy we find that the columns which contain the facts and theories relative to the *sun* and *planets,* after exhibiting several stages of induction within themselves, are at length suddenly connected with a column till then quite distinct, containing the *precession of the equinoxes.* In like manner, in the Table of Optics, the columns which contain the facts and theories relative to *double refraction,* and those which include *polarization by crystals,* each go separately through several stages of induction; and then these two sets of columns are suddenly connected by Fresnel's mathematical induction, that double refraction and polarization arise from the same cause: thus exhibiting a remarkable *Consilience.*

5. The constant *Tendency to Simplicity* in the sciences of which the progress is thus represented, appears from the form of the Table itself; for the single trunk into which all the branches converge, contains in itself the substance of all the propositions by means of which this last generalization was arrived at. It is true, that this ultimate result is sometimes not so simple as in the Table it appears: for instance, the ultimate generalization of the Table exhibiting the progress of Physical Optics,—namely, that Light consists in Undulations,—must be understood as including some other hypotheses; as, that the undulations are transverse, that the ether through which they are propagated has its elasticity in crystals and other transparent bodies regulated by certain laws; and the like. Yet still, even acknowledging all the complication thus implied, the Table in question evidences clearly enough the constant advance towards unity, consistency, and simplicity, which have marked the progress of this Theory. The same is the case in the Inductive Table of Astronomy in a still greater degree.

6. These Tables naturally afford the opportunity of assigning to

each of the distinct steps of which the progress of science consists, the name of the *Discoverer* to whom it is due. Every one of the inductive processes which the brackets of our Tables mark, directs our attention to some person by whom the induction was first distinctly made. These names I have endeavoured to put in their due places in the Tables; and the Inductive Tree of our knowledge in each science becomes, in this way, an exhibition of the claims of each discoverer to distinction, and, as it were, a Genealogical Tree of scientific nobility. It is by no means pretended that such a tree includes the names of all the meritorious labourers in each department of science. Many persons are most usefully employed in collecting and verifying truths, who do not advance to any new truths. The labours of a number of such are included in each stage of our ascent. But such Tables as we have now before us will present to us the names of all the most eminent discoverers: for the main steps of which the progress of science consists, are transitions from more particular to more general truths, and must therefore be rightly given by these Tables; and those must be the greatest names in science to whom the principal events of its advance are thus due.

7. The Tables, as we have presented them, exhibit the course by which we pass from Particular to General through various gradations, and so to the most general. They display the order of *discovery*. But by reading them in an inverted manner, beginning at the single comprehensive truths with which the Tables end, and tracing these back into the more partial truths, and these again into special facts, they answer another purpose;—they exhibit the process of *verification* of discoveries once made. For each of our general propositions is true in virtue of the truth of the narrower propositions which it involves; and we cannot satisfy ourselves of its truth in any other way than by ascertaining that these its constituent elements are true. To assure ourselves that the sun attracts the planets with forces varying inversely as the square of the distance, we must analyse by geometry the motion of a body in an ellipse about the focus, so as to see that such a motion does imply such a force. We must also verify those calculations by which the observed places of each planet are stated to be included in an ellipse. These calculations involve assumptions respecting the path which the earth describes about the sun, which assumptions must

again be verified by reference to observation. And thus, proceeding from step to step, we resolve the most general truths into their constituent parts; and these again into their parts; and by testing at each step, both the reality of the asserted ingredients and the propriety of the conjunction, we establish the whole system of truths, however wide and various it may be.

8. It is a very great advantage, in such a mode of exhibiting scientific truths, that it resolves the verification of the most complex and comprehensive theories, into a number of small steps, of which almost any one falls within the reach of common talents and industry. That *if* the particulars of any one step be true, the generalization also is true, any person with a mind properly disciplined may satisfy himself by a little study. That each of these particular propositions *is* true, may be ascertained, by the same kind of attention, when this proposition is resolved into *its* constituent and more special propositions. And thus we may proceed, till the most general truth is broken up into small and manageable portions. Of these portions, each may appear by itself narrow and easy; and yet they are so woven together, by hypothesis and conjunction, that the truth of the parts necessarily assures us of the truth of the whole. The verification is of the same nature as the verification of a large and complex statement of great sums received by a mercantile office on various accounts from many quarters. The statement is separated into certain comprehensive heads, and these into others less extensive; and these again into smaller collections of separate articles, each of which can be inquired into and reported on by separate persons. And thus at last, the mere addition of numbers performed by these various persons, and the summation of the results which they obtain, executed by other accountants, is a complete and entire security that there is no errour in the whole of the process.

9. This comparison of the process by which we verify scientific truth to the process of Book-keeping in a large commercial establishment, may appear to some persons not sufficiently dignified for the subject. But, in fact, the possibility of giving this formal and business-like aspect to the evidence of science, as involved in the process of successive generalization, is an inestimable advantage. For if no one could pronounce concerning a wide and profound theory except he

who could at once embrace in his mind the whole range of inference, extending from the special facts up to the most general principles, none but the greatest geniuses would be entitled to judge concerning the truth or errour of scientific discoveries. But, in reality, we seldom need to verify more than one or two steps of such discoveries at one time; and this may commonly be done (when the discoveries have been fully established and developed,) by any one who brings to the task clear conceptions and steady attention. The progress of science is gradual: the discoveries which are successively made, are also verified successively. We have never any very large collections of them on our hands at once. The doubts and uncertainties on any one who has studied science with care and perseverance are generally confined to a few points. If he can satisfy himself upon these, he has no misgivings respecting the rest of the structure; which has indeed been repeatedly verified by other persons in like manner. The fact that science is capable of being resolved into separate processes of verification, is that which renders it possible to form a great body of scientific truth, by adding together a vast number of truths of which many men, at various times and by multiplied efforts, have satisfied themselves. The treasury of Science is constantly rich and abundant, because it accumulates the wealth which is thus gathered by so many, and reckoned over by so many more: and the dignity of Knowledge is no more lowered by the multiplicity of the tasks on which her servants are employed, and the narrow field of labour to which some confine themselves, than the rich merchant is degraded by the number of offices which it is necessary for him to maintain, and the minute articles of which he requires an exact statement from his accountants.

10. The analysis of doctrines inductively obtained, into their constituent facts, and the arrangement of them in such a form that the conclusiveness of the induction may be distinctly seen, may be termed *the Logic of Induction*. By *Logic* has generally been meant a system which teaches us so to arrange our reasonings that their truth or falsehood shall be evident in their form. In *deductive* reasonings, in which the general principles are assumed, and the question is concerning their application and combination in particular cases, the device which thus enables us to judge whether our reasonings are conclusive is the *Syllogism;* and this *form,* along with the rules which be-

long to it, does in fact supply us with a criterion of deductive or demonstrative reasoning. The *Inductive Table,* such as it is presented in the present chapter, in like manner supplies the means of ascertaining the truth of our *inductive* inferences, so far as the *form* in which our reasoning may be stated can afford such a criterion. Of course some care is requisite in order to reduce a train of demonstration into the form of a series of syllogisms; and certainly not less thought and attention are required for resolving all the main doctrines of any great department of science into a graduated table of co-ordinate and subordinate inductions. But in each case, when this task is once executed, the evidence or want of evidence of our conclusions appears immediately in a most luminous manner. In each step of induction, our Table enumerates the particular facts, and states the general theoretical truth which includes these and which these constitute. The special act of attention by which we satisfy ourselves that the facts *are* so included,—that the general truth *is* so constituted,—then affords little room for errour, with moderate attention and clearness of thought.

11. We may find an example of this *act of attention* thus required, at any one of the steps of induction in our Tables; for instance, at the step in the early progress of astronomy at which it was inferred, that the earth is a globe, and that the sphere of the heavens (relatively) performs a diurnal revolution round this globe of the earth. How was this established in the belief of the Greeks, and how is it fixed in our conviction? As to the globular form, we find that as we travel to the north, the apparent pole of the heavenly motions, and the constellations which are near it, seem to mount higher, and as we proceed southwards they descend. Again, if we proceed from two different points considerably to the east and west of each other, and travel directly northwards from each, as from the south of Spain to the north of Scotland, and from Greece to Scandinavia, these two north and south lines will be much nearer to each other in their northern than in their southern parts. These and similar facts, as soon as they are clearly estimated and connected in the mind, are *seen to be consistent* with a convex surface of the earth, and with no other: and this notion is further confirmed by observing that the boundary of the earth's shadow upon the moon is always circular; it being supposed to be al-

ready established that the moon receives her light from the sun, and that lunar eclipses are caused by the interposition of the earth. As for the assertion of the (relative) diurnal revolution of the starry sphere, it is merely putting the visible phenomena in an exact geometrical form: and thus we establish and verify the doctrine of the revolution of the sphere of the heavens about the globe of the earth, by contemplating it so as to *see* that it does really and exactly include the particular facts from which it is collected.

We may, in like manner, illustrate this mode of verification by any of the other steps of the same Table. Thus if we take the great Induction of Copernicus, the heliocentric scheme of the solar system, we find it in the Table exhibited as including and explaining, *first*, the diurnal revolution just spoken of; *second*, the motions of the moon among the fixed stars; *third*, the motions of the planets with reference to the fixed stars and the sun; *fourth*, the motion of the sun in the ecliptic. And the scheme being clearly conceived, we *see* that all the particular facts *are* faithfully represented by it; and this agreement, along with the simplicity of the scheme, in which respect it is so far superior to any other conception of the solar system, persuade us that it is really the plan of nature.

In exactly the same way, if we attend to any of the several remarkable discoveries of Newton, which form the principal steps in the latter part of the Table, as for instance, the proposition that the sun attracts all the planets with a force which varies inversely as the square of the distance, we find it proved by its including three other propositions previously established;—*first*, that the sun's mean force on different planets follows the specified variation (which is proved from Kepler's third law); *second*, that the force by which each planet is acted upon in different parts of its orbit tends to the sun (which is proved by the equable description of areas); *third*, that this force in different parts of the same orbit is also inversely as the square of the distance (which is proved from the elliptical form of the orbit). And the Newtonian generalization, when its consequences are mathematically traced, is *seen* to agree with each of these particular propositions, and thus is fully established.

12. But when we say that the more general proposition *includes* the several more particular ones, we must recollect what has before

been said, that these particulars form the general truth, not by being merely enumerated and added together, but by being seen *in a new light*. No mere verbal recitation of the particulars can decide whether the general proposition is true; a special act of thought is requisite in order to determine how truly each is included in the supposed induction. In this respect the Inductive Table is not like a mere schedule of accounts, where the rightness of each part of the reckoning is tested by mere addition of the particulars. On the contrary, the Inductive truth is never the mere *sum* of the facts. It is made into something more by the introduction of a new mental element; and the mind, in order to be able to supply this element, must have peculiar endowments and discipline. Thus looking back at the instances noticed in the last article, how are we to see that a convex surface of the earth is necessarily implied by the convergence of meridians towards the north, or by the visible descent of the north pole of the heavens as we travel south? Manifestly the student, in order to see this, must have clear conceptions of the relations of space, either naturally inherent in his mind, or established there by geometrical cultivation,—by studying the properties of circles and spheres. When he is so prepared, he will feel the force of the expressions we have used, that the facts just mentioned are *seen to be consistent* with a globular form of the earth; but without such aptitude he will not see this consistency: and if this be so, the mere assertion of it in words will not avail him in satisfying himself of the truth of the proposition.

In like manner, in order to perceive the force of the Copernican induction, the student must have his mind so disciplined by geometrical studies, or otherwise, that he sees clearly how absolute motion and relative motion would alike produce apparent motion. He must have learnt to cast away all prejudices arising from the seeming fixity of the earth; and then he will see that there is nothing which stands in the way of the induction, while there is much which is on its side. And in the same manner the Newtonian induction of the law of the sun's force from the elliptical form of the orbit, will be evidently satisfactory to him only who has such an insight into Mechanics as to see that a curvilinear path must arise from a constantly deflecting force; and who is able to follow the steps of geometrical reasoning by which, from the properties of the ellipse, Newton proves this deflection to be in the proportion in which he asserts the force to be. And thus in all cases the inductive truth must indeed be verified by comparing it with the particular facts; but then this comparison is possible for him only whose mind is properly disciplined and prepared in the use of those conceptions, which, in addition to the facts, the act of induction requires.

13. In the Tables some indication is given, at several of the steps, of the act which the mind must thus perform, besides the mere conjunction of facts, in order to attain to the inductive truth. Thus in the cases of the Newtonian inductions just spoken of, the inferences are stated to be made 'By Mechanics;' and in the case of the Copernican induction, it is said that, 'By the nature of motion, the apparent motion is the same, whether the heavens or the earth have a diurnal motion; and the latter is more simple.' But these verbal statements are to be understood as mere hints:[19] they cannot supersede the necessity of the student's contemplating for himself the mechanical principles and the nature of motion thus referred to.

14. In the common or Syllogistic Logic, a certain *Formula* of language is used in stating the reasoning, and is useful in enabling us more readily to apply the Criterion of Form to alleged demonstrations. This formula is the usual Syllogism; with its members, Major Premiss, Minor Premiss, and Conclusion. It may naturally be asked whether in Inductive Logic there is any such Formula? whether there is any standard form of words in which we may most properly express the inference of a general truth from particular facts?

At first it might be supposed that the formula of Inductive Logic need only be of this kind: 'These particulars, and all known particulars of the same kind, are exactly included in the following general proposition.' But a moment's reflection on what has just been said will show us that this is not sufficient: for the particulars are not merely *included* in the general proposition. It is not enough that they appertain to it by enumeration. It is, for instance, no adequate example of Induction to say, 'Mercury describes an elliptical path, so does Venus, so do the Earth, Mars, Jupiter, Saturn, Uranus; therefore all the Planets describe elliptical paths.' This is, as we have seen, the mode of stating the *evidence* when the proposition is once suggested; but the Inductive step consists in the *suggestion* of a conception not before apparent. When Kepler, after trying to connect the observed places of the planet Mars in many other ways, found at last that the conception of an *ellipse* would include them all, he obtained a truth by induction: for this conclusion was not obviously included in the phenomena, and had not been applied to these facts previously. Thus in our Formula, besides stating that the particulars are included in the general proposition, we must also imply that the generality is constitued by a new Conception,—new at least in its application.

Hence our Inductive Formula might be something like the following: 'These particulars, and all known particulars of the same kind, are exactly expressed by adopting the Conceptions and Statement of the following Proposition.' It is of course requisite that the Conceptions should be perfectly clear, and should precisely embrace the facts, according to the explanation we have already given of those conditions.

15. It may happen, as we have already stated, that the Explication of a Conception, by which it acquires its due distinctness, leads to a Definition, which Definition may be taken as the summary and total result of the intellectual efforts to which this distinctness is due. In such cases, the Formula of Induction may be modified according to this condition; and we may state the inference by saying, after an enumeration and analysis of the appropriate facts, 'These facts are completely and distinctly expressed by adopting the following Definition and Proposition.'

This Formula has been adopted in stating the Inductive Propositions which constitute the basis of the science of Mechanics, in a work intitled *The Mechanical Euclid.* The fundamental truths of the subject are expressed in *Inductive Pairs* of Assertions, consisting each of a Definition and a Proposition, such as the following:

DEF.—A *Uniform Force* is that which acting in the direction of the body's motion, adds or subtracts equal velocities in equal times.

PROP.—Gravity is a Uniform Force.

Again,

DEF.—Two *Motions* are *compounded* when each produces its separate effect in a direction parallel to itself.

PROP.—When any Force acts upon a body in motion, the motion which the Force would produce in the body at rest is compounded with the previous motion of the body.

And in like manner in other cases.

In these cases the proposition is, of course, established, and the definition realized, by an enumeration of the facts. And in the case of inferences made in such a form, the Definition of the Conception and the Assertion of the Truth are both requisite and are correlative to one another. Each of the two steps contains the verification and justification of the other. The Proposition derives its meaning from the Definition; the Definition derives its reality from the Proposition. If they are separated, the Definition is arbitrary or empty, the Proposition vague or ambiguous.

16. But it must be observed that neither of the preceding Formulae expresses the full cogency of the inductive proof. They declare only that the results *can* be clearly explained and rigorously deduced by the employment of a certain Definition and a certain Proposition. But in order to make the conclusion demonstrative, which in perfect examples of Induction it is, we ought to be able to declare that the results can be clearly explained and rigorously declared *only* by the Definition and Proposition which we adopt. And in reality, the conviction of the sound inductive reasoner does reach to this point. The Mathematician asserts the Laws of Motion, seeing clearly that they (or laws equivalent to them) afford the only means of clearly expressing and deducing the actual facts. But this conviction, that the inductive inference is not only consistent with the facts, but necessary, finds its place in the mind gradually, as the contemplation of the consequences of the proposition, and the various relations of the facts, becomes steady and familiar. It is scarcely possible for the student at once to satisfy himself that the inference is thus inevitable. And when he arrives at this conviction, he sees also, in many cases at least, that there may be other ways of expressing the substance of the truth established, besides that special Proposition which he has under his notice.

We may, therefore, without impropriety, renounce the undertaking of conveying in our formula this final conviction of the necessary truth of our inference. We may leave it to be thought, without insisting upon saying it, that in such cases what *can* be true, *is* true. But if we wish to express the ultimate significance of the Inductive Act of thought, we may take as our Formula for the Colligation of

Facts by Induction, this:—'The several Facts are exactly expressed as one Fact if, *and only if,* we adopt the Conception and the Assertion' of the inductive inference.

17. I have said that the mind must be properly disciplined in order that it may see the necessary connexion between the facts and the general proposition in which they are included. And the perception of this connexion, though treated as *one step* in our inductive inference, may imply *many steps* of demonstrative proof. The connexion is this, that the particular case is included in the general one, that is, may be *deduced* from it: but this deduction may often require many links of reasoning. Thus in the case of the inference of the law of the force from the elliptical form of the orbit by Newton, the proof that in the ellipse the deflection from the tangent is inversely as the square of the distance from the focus of the ellipse, is a ratiocination consisting of several steps, and involving several properties of Conic Sections; these properties being supposed to be previously established by a geometrical system of demonstration on the special subject of the Conic Sections. In this and similar cases the Induction involves many steps of Deduction. And in such cases, although the Inductive Step, the Invention of the Conception, is really the most important, yet since, when once made, it occupies a familiar place in men's minds; and since the Deductive Demonstration is of considerable length and requires intellectual effort to follow it at every step, men often admire the deductive part of the proposition, the geometrical or algebraical demonstration, far more than that part in which the philosophical merit really resides.

18. Deductive reasoning is virtually a collection of syllogisms, as has already been stated; and in such reasoning, the general principles, the Definitions and Axioms, necessarily stand at the *beginning* of the demonstration. In an inductive inference, the Definitions and Principles are the *final result* of the reasoning, the ultimate effect of the proof. Hence when an Inductive Proposition is to be established by a proof involving several steps of demonstrative reasoning, the enunciation of the Proposition will contain, explicitly or implicitly, principles which the demonstration proceeds upon as axioms, but which are really inductive inferences. Thus in order to prove that the force which retains a planet in an ellipse varies inversely as the square of the dis-

tance, it is taken for granted that the Laws of Motion are true, and that they apply to the planets. Yet the doctrine that this is so, as well as the law of the force, were established only by this and the like demonstrations. The Doctrine which is the hypothesis of the deductive reasoning, is the *inference* of the inductive process. The special facts which are the basis of the inductive inference, are the conclusion of the train of deduction. And in this manner the deduction establishes the induction. The principle which we gather from the facts is true, because the facts can be derived from it by rigorous demonstration. Induction moves upwards, and deduction downwards, on the same stair.

But still there is a great difference in the character of their movements. Deduction descends steadily and methodically, step by step: Induction mounts by a leap which is out of the reach of method. She bounds to the top of the stair at once; and then it is the business of Deduction, by trying each step in order, to establish the solidity of her companion's footing. Yet these must be processes of the same mind. The Inductive Intellect makes an assertion which is subsequently justified by demonstration; and it shows its sagacity, its peculiar character, by enunciating the proposition when as yet the demonstration does not exist: but then it shows that it *is* sagacity, by also producing the demonstration.

It has been said that inductive and deductive reasoning are contrary in their scheme; that in Deduction we infer particular from general truths; while in Induction we infer general from particular: that Deduction consists of many steps, in each of which we apply known general propositions in particular cases; while in Induction we have a single step, in which we pass from many particular truths to one general proposition. And this is truly said; but though contrary in their motions, the two are the operation of the same mind travelling over the same ground. Deduction is a necessary part of Induction. Deduction justifies by calculation what Induction had happily guessed. Induction recognizes the ore of truth by its weight; Deduction confirms the recognition by chemical analysis. Every step of Induction must be confirmed by rigorous deductive reasoning, followed into such detail as the nature and complexity of the relations (whether of quantity or any other) render requisite. If not so justified by the supposed discoverer, it is *not* Induction.

19. Such Tabular arrangements of propositions as we have constructed may be considered as the *Criterion of Truth* for the doctrines which they include. They are the Criterion of Inductive Truth, in the same sense in which Syllogistic Demonstration is the Criterion of Necessary Truth,—of the certainty of conclusions, depending upon evident First Principles. And that such Tables are really a Criterion of the truth of the propositions which they contain, will be plain by examining their structure. For if the connexion which the inductive process assumes be ascertained to be in each case real and true, the assertion of the general proposition merely collects together ascertained truths; and in like manner each of those more particular propositions is true, because it merely expresses collectively more special facts: so that the most general theory is only the assertion of a great body of facts, duly classified and subordinated. When we assert the truth of the Copernican theory of the motions of the solar system, or of the Newtonian theory of the forces by which they are caused, we merely assert the groups of propositions which, in the Table of Astronomical Induction, are included in these doctrines; and ultimately, we may consider ourselves as merely asserting at once so many Facts, and therefore, of course, expressing an indisputable truth.

20. At any one of these steps of Induction in the Table, the inductive proposition is a *Theory* with regard to the Facts which it includes, while it is to be looked upon as a *Fact* with respect to the higher generalizations in which it is included. In any other sense, as was formerly shown, the opposition of *Fact* and *Theory* is untenable, and leads to endless perplexity and debate. Is it a Fact or a Theory that the planet Mars revolves in an Ellipse about the Sun? To Kepler, employed in endeavouring to combine the separate observations by the Conception of an Ellipse, it is a Theory; to Newton, engaged in inferring the law of force from a knowledge of the elliptical motion, it is a Fact. There are, as we have already seen, no special attributes of Theory and Fact which distinquish them from one another. Facts are phenomena apprehended by the aid of conceptions and mental acts, as Theories also are. We commonly call our observations *Facts,* when we apply, without effort or consciousness, conceptions perfectly fa-

miliar to us: while we speak of Theories, when we have previously contemplated the Facts and the connecting Conception separately, and have made the connexion by a conscious mental act. The real difference is a difference of relation; as the same proposition in a demonstration is the *premiss* of one syllogism and the *conclusion* in another;—as the same person is a father and a son. Propositions are Facts and Theories, according as they stand above or below the Inductive Brackets of our Tables.

21. To obviate mistakes I may remark that the terms *higher* and *lower*, when used of generalizations, are unavoidably represented by their opposites in our Inductive Tables. The highest generalization is that which includes all others; and this stands the lowest on our page, because, reading downwards, that is the place which we last reach.

There is a distinction of the knowledge acquired by Scientific Induction into two kinds, which is so important that we shall consider it in the succeeding chapter.

## CHAPTER VII
## OF LAWS OF PHENOMENA AND OF CAUSES

*Aphorism XXIV*

*Inductive truths are of two kinds,* Laws of Phenomena, *and* Theories of Causes. *It is necessary to begin in every science with the Laws of Phenomena; but it is impossible that we should be satisfied to stop short of a Theory of Causes. In Physical Astronomy, Physical Optics, Geology, and other sciences, we have instances showing that we can make a great advance in inquiries after true Theories of Causes.*

1. In the first attempts at acquiring an exact and connected knowledge of the appearances and operations which nature presents, men went no further than to learn *what* takes place, not *why* it occurs. They discovered an Order which the phenomena follow, Rules which they obey; but they did not come in sight of the Powers by which these rules are determined, the Causes of which this order is the effect. Thus, for example, they found that many of the celestial motions took place as if the sun and stars were carried round by the revolutions of certain celestial spheres; but what causes kept these

spheres in constant motion, they were never able to explain. In like manner in modern times, Kepler discovered that the planets describe ellipses, before Newton explained why they select this particular curve, and describe it in a particular manner. The laws of reflection, refraction, dispersion, and other properties of light have long been known; the causes of these laws are at present under discussion. And the same might be said of many other sciences. The discovery of *the Laws of Phenomena* is, in all cases, the first step in exact knowledge; these Laws may often for a long period constitute the whole of our science; and it is always a matter requiring great talents and great efforts, to advance to a knowledge of the *Causes* of the phenomena.

Hence the larger part of our knowledge of nature, at least of the certain portion of it, consists of the knowledge of the Laws of Phenomena. In Astronomy indeed, besides knowing the rules which guide the appearances, and resolving them into the real motions from which they arise, we can refer these motions to the forces which produce them. In Optics, we have become acquainted with a vast number of laws by which varied and beautiful phenomena are governed; and perhaps we may assume, since the evidence of the Undulatory Theory has been so fully developed, that we know also the Causes of the Phenomena. But in a large class of sciences, while we have learnt many Laws of Phenomena, the causes by which these are produced are still unknown or disputed. Are we to ascribe to the operation of a fluid or fluids, and if so, in what manner, the facts of heat, magnetism, electricity, galvanism? What are the forces, by which the elements of chemical compounds are held together? What are the forces, of a higher order, as we cannot help believing, by which the course of vital action in organized bodies is kept up? In these and other cases, we have extensive departments of science; but we are as yet unable to trace the effects to their causes; and our science, so far as it is positive and certain, consists entirely of the laws of phenomena.

2. In those cases in which we have a division of the science which teaches us the doctrine of the causes, as well as one which states the rules which the effects follow, I have, in the *History*, distinguished the two portions of the science by certain terms. I have thus spoken of *Formal* Astronomy and *Physical* Astronomy. The latter phrase has long been commonly employed to describe that department of Astron-

omy which deals with those forces by which the heavenly bodies are guided in their motions; the former adjective appears well suited to describe a collection of rules depending on those ideas of space, time, position, number, which are, as we have already said, the *forms* of our apprehension of phenomena. The laws of phenomena may be considered as *formulae,* expressing results in terms of those ideas. In like manner, I have spoken of Formal Optics and Physical Optics; the latter division including all speculations concerning the machinery by which the effects are produced. Formal Acoustics and Physical Acoustics may be distinguished in like manner, although these two portions of science have been a good deal mixed together by most of those who have treated of them. Formal Thermotics, the knowledge of the laws of the phenomena of heat, ought in like manner to lead to Physical Thermotics, or the Theory of Heat with reference to the cause by which its effects are produced;—a branch of science which as yet can hardly be said to exist.

3. What *kinds of cause* are we to admit in science? This is an important, and by no means as easy question. In order to answer it, we must consider in what manner our progress in the knowledge of causes has hitherto been made. By far the most conspicuous instance of success in such researches, is the discovery of the causes of the motions of the heavenly bodies. In this case, after the formal laws of the motions, —their conditions as to space and time,—had become known, men were enabled to go a step further; to reduce them to the familiar and general cause of motion—mechanical force; and to determine the laws which this force follows. That this was a step in addition to the knowledge previously possessed, and that it was a real and peculiar truth, will not be contested. And a step in any other subject which should be analogous to this in astronomy;—a discovery of causes and forces as certain and clear as the discovery of universal gravitation;—would undoubtedly be a vast advance upon a body of science consisting only of the laws of phenomena.

4. But although physical astronomy may well be taken as a standard in estimating the value and magnitude of the advance from the knowledge of phenomena to the knowledge of causes; the peculiar features of the transition from formal to physical science in that subject must not be allowed to limit too narrowly our views of the nature

of this transition in other cases. We are not, for example, to consider that the step which leads us to the knowledge of causes in any province of nature must necessarily consist in the discovery of centers of forces, and collections of such centers, by which the effects are produced. The discovery of the causes of phenomena may imply the detection of a fluid by whose undulations, or other operations, the results are occasioned. The phenomena of acoustics are, we know, produced in this manner by the air; and in the cases of light, heat, magnetism, and others, even if we reject all the theories of such fluids which have hitherto been proposed, we still cannot deny that such theories are intelligible and possible, as the discussions concerning them have shown. Nor can it be doubted that if the assumption of such a fluid, in any case, were as well evidenced as the doctrine of universal gravitation is, it must be considered as a highly valuable theory.

5. But again; not only must we, in aiming at the formation of a Causal Section in each Science of Phenomena, consider Fluids and their various modes of operation admissible, as well as centers of mechanical force; but we must be prepared, if it be necessary, to consider the forces, or powers to which we refer the phenomena, under still more general aspects, and invested with characters different from mere mechanical force. For example; the forces by which the chemical elements of bodies are bound together, and from which arise, both their sensible texture, their crystalline form, and their chemical composition, are certainly forces of a very different nature from the mere attraction of matter according to its mass. The powers of assimilation and reproduction in plants and animals are obviously still more removed from mere mechanism; yet these powers are not on that account less real, nor a less fit and worthy subject of scientific inquiry.

6. In fact, these forces—mechanical, chemical and vital,—as we advance from one to the other, each bring into our consideration new characters; and what these characters are, has appeared in the historical survey which we made of the Fundamental Ideas of the various sciences. It was then shown that the forces by which chemical effects are produced necessarily involve the Idea of Polarity,—they are polar forces; the particles tend together in virtue of opposite properties which in the combination neutralize each other. Hence, in attempting to advance to a theory of Causes in chemistry, our task is by no means

to invent laws of *mechanical* force, and collections of forces, by which the effects may be produced. We know beforehand that no such attempt can succeed. Our aim must be to conceive such new kinds of force, including Polarity among their characters, as may best render the results intelligible.

7. Thus in advancing to a Science of Cause in any subject, the labour and the struggle is, not to analyse the phenomena according to any preconceived and already familiar ideas, but to form distinctly new conceptions, such as do really carry us to a more intimate view of the processes of nature. Thus in the case of astronomy, the obstacle which deferred the discovery of the true causes from the time of Kepler to that of Newton, was the difficulty of taking hold of mechanical conceptions and axioms with sufficient clearness and steadiness; which, during the whole of that interval, mathematicians were learning to do. In the question of causation which now lies most immediately in the path of science, that of the causes of electrical and chemical phenomena, the business of rightly fixing and limiting the conception of polarity, is the proper object of the efforts of discoverers. Accordingly a large portion of Mr Faraday's recent labours is directed, not to the attempt at discovering new laws of phenomena, but to the task of throwing light upon the conception of polarity, and of showing how it must be understood, so that it shall include electrical induction and other phenomena, which have commonly been ascribed to forces acting mechanically at a distance. He is by no means content, nor would it answer the ends of science that he should be, with stating the results of his experiments; he is constantly, in every page, pointing out the interpretation of his experiments, and showing how the conception of Polar Forces enters into this interpretation. 'I shall,' he says, 'use every opportunity which presents itself of returning to that strong test of truth, experiment; but,' he adds, 'I shall necessarily have occasion to speak theoretically, and even hypothetically.' His hypothesis that electrical inductive action always takes place by means of a continuous line of polarized particles, and not by attraction and repulsion at a distance, if established, cannot fail to be a great step on our way towards a knowledge of causes, as well as phenomena, in the subjects under his consideration.

8. The process of obtaining new conceptions is, to most minds, far

more unwelcome than any labour in employing old ideas. The effort is indeed painful and oppressive; it is feeling in the dark for an object which we cannot find. Hence it is not surprising that we should far more willingly proceed to seek for new causes by applying conceptions borrowed from old ones. Men were familiar with solid frames, and with whirlpools of fluid, when they had not learnt to form any clear conception of attraction at a distance. Hence they at first imagined the heavenly motions to be caused by Crystalline Spheres, and by Vortices. At length they were taught to conceive Central Forces, and then they reduced the solar system to these. But having done this, they fancied that all the rest of the machinery of nature must be central forces. We find Newton expressing this conviction, and the mathematicians of the last century acted upon it very extensively. We may especially remark Laplace's labours in this field. Having explained, by such forces, the phenomena of capillary attraction, he attempted to apply the same kind of explanation to the reflection, refraction, and double refraction of light;—to the constitution of gases;—to the operation of heat. It was soon seen that the explanation of refraction was arbitrary, and that of double refraction illusory; while polarization entirely eluded the grasp of this machinery. Centers of force would no longer represent the modes of causation which belonged to the phenomena. Polarization required some other contrivance, such as the undulatory theory supplied. No theory of light can be of any avail in which the fundamental idea of Polarity is not clearly exhibited.

9. The sciences of magnetism and electricity have given rise to theories in which this relation of polarity is exhibited by means of two opposite fluids;[20]—a positive and a negative fluid, or a vitreous and a resinous, for electricity, and a boreal and an austral fluid for magnetism. The hypothesis of such fluids gives results agreeing in a remarkable manner with the facts and their measures, as Coulomb and others have shown. It may be asked how far we may, in such a case, suppose that we have discovered the true cause of the phenomena, and whether it is sufficiently proved that these fluids really exist. The right answer seems to be, that the hypothesis certainly represents the truth so far as regards the polar relation of the two energies, and the laws of the attractive and repulsive forces of the particles in which

these energies reside; but that we are not entitled to assume that the vehicles of these energies possess other attributes of material fluids, or that the forces thus ascribed to the particles are the primary elementary forces from which the action originates. We are the more bound to place this cautious limit to our acceptance of the Coulombian theory, since in electricity Faraday has in vain endeavoured to bring into view one of the polar fluids without the other: whereas such a result ought to be possible if there were two separable fluids. The impossibility of this separate exhibition of one fluid appears to show that the fluids are *real* only so far as they are *polar*. And Faraday's view above mentioned, according to which the attractions at a distance are resolved into the action of lines of polarized particles of air, appears still further to show that the conceptions hitherto entertained of electrical forces, according to the Coulombian theory, do not penetrate to the real and intimate nature of the causation belonging to this case.

10. Since it is thus difficult to know when we have seized the true cause of the phenomena in any department of science, it may appear to some persons that physical inquirers are imprudent and unphilosophical in undertaking this Research of Causes; and that it would be safer and wiser to confine ourselves to the investigation of the laws of phenomena, in which field the knowledge which we obtain is definite and certain. Hence there have not been wanting those who have laid it down as a maxim that 'science must study only the laws of phenomena, and never the mode of production.'[21] But it is easy to see that such a maxim would confine the breadth and depth of scientific inquiries to a most scanty and miserable limit. Indeed, such a rule would defeat its own object; for the laws of phenomena, in many cases, cannot be even expressed or understood without some hypothesis respecting their mode of production. How could the phenomena of polarization have been conceived or reasoned upon, except by imagining a polar arrangement of particles, or transverse vibrations, or some equivalent hypothesis? The doctrines of fits of easy transmission, the doctrine of moveable polarization, and the like, even when erroneous as representing the whole of the phenomena, were still useful in combining some of them into laws; and without some such hypotheses the facts could not have been followed out. The doctrine

of a fluid caloric may be false; but without imagining such a fluid, how could the movement of heat from one part of a body to another be conceived? It may be replied that Fourier, Laplace, Poisson, who have principally cultivated the Theory of Heat, have not conceived it as a fluid, but have referred conduction to the radiation of the molecules of bodies, which they suppose to be separate points. But this molecular constitution of bodies is itself an assumption of the mode in which the phenomena are produced; and the radiation of heat suggests inquiries concerning a fluid emanation, no less than its conduction does. In like manner, the attempts to connect the laws of phenomena of heat and of gases, have led to hypotheses respecting the constitution of gases, and the combination of their particles with those of caloric, which hypotheses may be false, but are probably the best means of discovering the truth.

To debar science from inquiries like these, on the ground that it is her business to inquire into facts, and not to speculate about causes, is a curious example of that barren caution which hopes for truth without daring to venture upon the quest of it. This temper would have stopped with Kepler's discoveries, and would have refused to go on with Newton to inquire into the mode in which the phenomena are produced. It would have stopped with Newton's optical facts, and would have refused to go on with him and his successors to inquire into the mode in which these phenomena are produced. And, as we have abundantly shown, it would, on that very account, have failed in seeing what the phenomena really are.

In many subjects the attempt to study the laws of phenomena, independently of any speculations respecting the causes which have produced them, is neither possible for human intelligence nor for human temper. Men cannot contemplate the phenomena without clothing them in terms of some hypothesis, and will not be schooled to suppress the questionings which at every moment rise up within them concerning the causes of the phenomena. Who can attend to the appearances which come under the notice of the geologist;—strata regularly bedded, full of the remains of animals such as now live in the depths of the ocean, raised to the tops of mountains, broken, contorted, mixed with rocks such as still flow from the mouths of volcanos;—who can see phenomena like these, and imagine that he best promotes the progress of our knowledge of the earth's history, by noting down the facts, and abstaining from all inquiry whether these are really proofs of past states of the earth and of subterraneous forces, or merely an accidental imitation of the effects of such causes? In this and similar cases, to proscribe the inquiry into causes would be to annihilate the science.

Finally, this caution does not even gain its own single end, the escape from hypotheses. For, as we have said, those who will not seek for new and appropriate causes of newly-studied phenomena, are almost inevitably led to ascribe the facts to modifications of causes already familiar. They may declare that they will not hear of such causes as vital powers, elective affinities, electric, or calorific, or luminiferous ethers or fluids; but they will not the less on that account assume hypotheses equally unauthorized;—for instance—universal mechanical forces; a molecular constitution of bodies; solid, hard, inert matter;—and will apply these hypotheses in a manner which is arbitrary in itself as well as quite insufficient for its purpose.

11. It appears, then, to be required, both by the analogy of the most successful efforts of science in past times and by the irrepressible speculative powers of the human mind, that we should attempt to discover both the *laws of phenomena,* and their *causes.* In every department of science, when prosecuted far enough, these two great steps of investigation must succeed each other. The laws of phenomena must be known before we can speculate concerning causes; the causes must be inquired into when the phenomena have been reduced to rule. In both these speculations the suppositions and conceptions which occur must be constantly tested by reference to observation and experiment. In both we must, as far as possible, devise hypotheses which, when we thus test them, display those characters of truth of which we have already spoken;—an agreement with facts such as will stand the most patient and rigid inquiry; a provision for predicting truly the results of untried cases; a consilience of inductions from various classes of facts; and a progressive tendancy of the scheme to simplicity and unity....

## CHAPTER VIII
## OF ART AND SCIENCE

***Aphorism XXV***

*Art and Science differ. The object of Science is Knowledge; the objects of Art, are Works. In Art, truth is a means to an end; in Science, it is the only end. Hence the Practical Arts are not to be classed among the Sciences.*

***Aphorism XXVI***

*Practical Knowledge, such as Art implies, is not Knowledge such as Science includes. Brute animals have a practical knowledge of relations of space and force; but they have no knowledge of Geometry or Mechanics....*

## CHAPTER IX
## OF THE CLASSIFICATION OF SCIENCES

1. The Classification of Sciences has its chief use in pointing out to us the extent of our powers of arriving at truth, and the analogies which may obtain between those certain and lucid portions of knowledge with which we are here concerned, and those other portions, of a very different interest and evidence, which we here purposely abstain to touch upon. The classification of human knowledge will, therefore, have a more peculiar importance when we can include in it the moral, political, and metaphysical, as well as the physical portions of our knowledge. But such a survey does not belong to our present undertaking: and a general view of the connexion and order of the branches of sciences which our review has hitherto included, will even now possess some interest; and may serve hereafter as an introduction to a more complete scheme of the general body of human knowledge.

2. In this, as in any other case, a sound classification must be the result, not of any assumed principles imperatively applied to the subject, but of an examination of the objects to be classified;—of an analysis of them into the principles in which they agree and differ. The Classification of Sciences must result from the consideration of their nature and contents. Accordingly, that review of the Sciences in which the *History* of the Sciences engaged us, led to a Classification, of which the main features are indicated in that work. The Classification thus obtained, depends neither upon the faculties of the mind to which the separate parts of our knowledge owe their origin, nor upon the objects which each science contemplates; but upon a more natural and

fundamental element;—namely, the *Ideas* which each science involves. The Ideas regulate and connect the facts, and are the foundations of the reasoning, in each science: and having in another work more fully examined these *Ideas,* we are now prepared to state here the classification to which they lead. If we have rightly traced each science to the Conceptions which are really fundamental *with regard to it,* and which give rise to the first principles on which it depends, it is not necessary for our purpose that we should decide whether these Conceptions are absolutely ultimate principles of thought, or whether, on the contrary, they can be further resolved into other Fundamental Ideas. We need not now suppose it determined whether or not *Number* is a mere modification of the Idea of Time, and *Force* a mere modification of the Idea of Cause: for however this may be, our Conception of Number is the foundation of Arithmetic, and our Conception of Force is the foundation of Mechanics. It is to be observed also that in our classification, each Science may involve, not only the Ideas or Conceptions which are placed opposite to it in the list, but also all which *precede* it. Thus Formal Astronomy involves not only the Conception of Motion, but also those which are the foundation of Arithmetic and Geometry. In like manner, Physical Astronomy employs the Sciences of Statics and Dynamics, and thus, rests on their foundations; and they, in turn, depend upon the Ideas of Space and of Time, as well as of Cause.

3. We further observe, that this arrangement of Sciences according to the Fundamental Ideas which they involve, points out the transition from those parts of human knowledge which have been included in our History and Philosophy, to other regions of speculation into which we have not entered. We have repeatedly found ourselves upon the borders of inquiries of a psychological, or moral, or theological nature. Thus the History of Physiology[22] led us to the consideration of Life, Sensation, and Volition; and at these Ideas we stopped, that we might not transgress the boundaries of our subject as then predetermined. It is plain that the pursuit of such conceptions and their consequences, would lead us to the sciences (if we are allowed to call them sciences) which contemplate not only animal, but human principles of action, to Anthropology, and Psychology. In other ways, too, the Ideas which we have examined, although manifestly the foundations of sciences

such as we have here treated of, also plainly pointed to speculations of a different order; thus the Idea of a Final Cause is an indispensable guide in Biology, as we have seen; but the conception of Design as directing the order of nature, once admitted, soon carries us to higher contemplations. Again, the Class of Palætiological Sciences which we were in the *History* led to construct, although we there admitted only one example of the Class, namely Geology, does in reality include many vast lines of research; as the history and causes of the diffusion of plants and animals, the history of languages, arts, and consequently of civilization. Along with these researches, comes the question how far these histories point backwards to a natural or a supernatural origin; and the Idea of a First Cause is thus brought under our consideration. Finally, it is not difficult to see that as the Physical Sciences have their peculiar governing Ideas, which support and shape them, so the Moral and Political Sciences also must similarly have their fundamental and formative Ideas, the source of universal and certain truths, each of their proper kind. But to follow out the traces of this analogy, and to verify the existence of those Fundamental Ideas in Morals and Politics, is a task quite out of the sphere of the work in which we are here engaged.

4. We may now place before the reader our Classification of the Sciences. I have added to the list of Sciences, a few not belonging to our present subject, that the nature of the transition by which we are to extend our philosophy into a wider and higher region may be in some measure perceived.

The Classification of the Sciences is given [on page 190].

A few remarks upon it offer themselves.

The *Pure* Mathematical Sciences can hardly be called *Inductive* Sciences. Their principles are not obtained by Induction from Facts, but are necessarily assumed in reasoning upon the subject matter which those sciences involve.

The Astronomy of the Ancients aimed only at explaining the motions of the heavenly bodies, as a *mechanism*. Modern Astronomy explains these motions on the principles of Mechanics.

The term Physics, when confined to a peculiar class of Sciences, is usually understood to exclude the Mechanical Sciences on the one side, and Chemistry on the other; and thus embraces the Secondary

Mechanical and Analytico-Mechanical Sciences. But the adjective *Physical* applied to any science and opposed to *Formal,* as in Astronomy and Optics, implies those speculations in which we consider not only the Laws of Phenomena but their Causes; and generally, as in those cases, their Mechanical Causes.

The term *Metaphysics* is applied to subjects in which the Facts examined are emotions, thoughts and mental conditions; subjects not included in our present survey.

CLASSIFICATION OF SCIENCES

| Fundamental Ideals or Conceptions | Sciences | Classification |
|---|---|---|
| Space | Geometry | Pure Mathematical Sciences |
| Time | | |
| *Number* | Arithmetic | |
| Sign | Algebra | |
| Limit | Differentials | |
| *Motion* | Pure Mechanism | Pure Motional Sciences |
| | Formal Astronomy | |
| Cause | | |
| *Force* | Statics | Mechanical Sciences |
| *Matter* | Dynamics | |
| *Inertia* | Hydrostatics | |
| *Fluid Pressure* | Hydrodynamics | |
| | Physical Astronomy | |
| Outness | | |
| Medium *of Sensation* | Acoustics | Secondary Mechanical Sciences (*Physics*) |
| Intensity *of Qualities* | Formal Optics | |
| *Scales of Qualities* | Physical Optics | |
| | Thermotics | |
| | Atmology | |
| Polarity | Electricity | Analytico-Mechanical Sciences (*Physics*) |
| | Magnetism | |
| | Galvanism | |
| Element (*Composition*) | | |
| *Chemical* Affinity | | |
| Substance (*Atoms*) | Chemistry | Analytical Science. |
| Symmetry | Crystallography | Analytico-Classificatory Sciences |
| Likeness | Systematic Mineralogy | |
| *Degrees of Likeness* | Systematic Botony | Classificatory Sciences |
| | Systematic Zoology | |
| *Natural* Affinity | Comparative Anatomy | |
| (*Vital Powers*) | | |
| Assimilation | | |
| Irritability | | |
| (*Organization*) | Biology | Organical Sciences |
| Final Cause | | |
| Instinct | | |
| Emotion | Psychology | (*Metaphysics*) |
| Thought | | |
| Historical Causation | Geology | Palætiological Sciences |
| | Distribution of Plants and Animals | |
| | Glossology | |
| | Ethnography | |
| First Cause | Natural Theology. | |

# NOVUM ORGANON RENOVATUM

## BOOK III

## OF METHODS EMPLOYED IN THE FORMATION OF SCIENCE

---

### CHAPTER I
### INTRODUCTION

***Aphorism XXVII***

*The Methods by which the construction of Science is promoted are,* Methods of Observation, Methods of obtaining clear Ideas, *and* Methods of Induction.

1. In the preceding Book, we pointed out certain general Characters of scientific knowledge which may often serve to distinguish it from opinions of a looser or vaguer kind. In the course of the progress of knowledge from the earliest to the present time, men have been led to a perception, more or less clear, of these characteristics. Various philosophers, from Plato and Aristotle in the ancient world, to Richard de Saint Victor and Roger Bacon in the middle ages, Galileo and Gilbert, Francis Bacon and Isaac Newton, in modern times, were led to offer precepts and maxims, as fitted to guide us to a real and fundamental knowledge of nature. It may on another occasion be our business to estimate the value of these precepts and maxims. And other contributions of the same kind to the philosophy of science might be noticed, and some which contain still more valuable suggestions, and indicate a more practical acquaintance with the subject. Among these, I must especially distinguish Sir John Herschel's *Discourse on the Study of Natural Philosophy*. But my object at present is not to relate the history, but to present the really valuable results of preceding labours: and I shall endeavour to collect, both from them and from my own researches

and reflections, such views and such rules as seem best adapted to assist us in the discovery and recognition of scientific truth; or, at least, such as may enable us to understand the process by which this truth is obtained. I would present to the reader the Philosophy and, if possible, the Art of Discovery.

2. But, in truth, we must acknowledge, before we proceed with this subject, that, speaking with strictness, an Art of *Discovery* is not possible;—that we can give no Rules for the pursuit of truth which shall be universally and peremptorily applicable;—and that the helps which we can offer to the inquirer in such cases are limited and precarious. Still, we trust it will be found that aids may be pointed out which are neither worthless nor uninstructive. The mere classification of examples of successful inquiry, to which our rules give occasion, is full of interest for the philosophical speculator. And if our maxims direct the discoverer to no operations which might not have occurred to his mind of themselves, they may still concentrate our attention on that which is most important and characteristic in these operations, and may direct us to the best mode of insuring their success. I shall, therefore, attempt to resolve the Process of Discovery into its parts, and to give an account as distinct as may be of Rules and Methods which belong to each portion of the process.

3. In Book II, we considered the three main parts of the process by which science is constructed: namely, the Decomposition and Observation of Complex Facts; the Explication of our Ideal Conceptions; and the Colligation of Elementary Facts by means of those Conceptions. The first and last of these three steps are capable of receiving additional accuracy by peculiar processes. They may further the advance of science in a more effectual manner, when directed by special technical *Methods,* of which in the present Book we must give a brief view. In this more technical form, the observation of facts involves the *Measurement of Phenomena;* and the Colligation of Facts includes all arts and rules by which the process of Induction can be assisted. Hence we shall have here to consider *Methods of Observation,* and *Methods of Induction,* using these phrases in the widest sense. The second of the three steps above mentioned, the Explication of our Conceptions, does not admit of being much assisted by methods, although something may be done by Education and Discussion.

4. The Methods of Induction, of which we have to speak, apply only to the first step in our ascent from phenomena to laws of nature;—the discovery of *Laws of Phenomena*. A higher and ulterior step remains behind, and follows in natural order the discovery of Laws of Phenomena; namely, the *Discovery of Causes;* and this must be stated as a distinct and essential process in a complete view of the course of science. Again, when we have thus ascended to the causes of phenomena and of their laws, we can often reason downwards from the cause so discovered; and we are thus led to suggestions of new phenomena, or to new explanations of phenomena already known. Such proceedings may be termed *Applications* of our Discoveries; including in the phrase, *Verifications* of our Doctirnes by such an application of them to observed facts. Hence we have the following series of processes concerned in the formation of science.

(1.) Decomposition of Facts;
(2.) Measurement of Phenomena;
(3.) Explication of Conceptions;
(4.) Induction of Laws of Phenomena;
(5.) Induction of Causes;
(6.) Application of Inductive Discoveries.

5. Of these six processes, the methods by which the second and fourth may be assisted are here our peculiar object of attention. The treatment of these subjects in the present work must necessarily be scanty and imperfect, although we may perhaps be able to add something to what has hitherto been systematically taught on these heads. Methods of Observation and of Induction might of themselves form an abundant subject for a treatise, and hereafter probably will do so, in the hands of future writers. A few remarks, offered as contributions to this subject, may serve to show how extensive it is, and how much more ready it now is than it ever before was, for a systematic discussion.

Of the above steps of the formation of science, the first, the Decomposition of Facts, has already been sufficiently explained in the last Book: for if we pursue it into further detail and exactitude, we find that we gradually trench upon some of the succeeding parts. I, therefore, proceed to treat of the second step, the Measurement of Phenomena;—of *Methods* by which this work, in its widest sense, is executed, and these I shall term Methods of Observation.

## CHAPTER II
## OF METHODS OF OBSERVATION

*Aphorism XXVIII*

*The Methods of Observation of Quantity in general are,* Numeration, *which is precise by the nature of Number; the* Measurement of Space *and* of Time, *which are easily made precise; the* Conversion of Space and Time, *by which each aids the measurement of the other; the* Method of Repetition; *the* Method of Coincidences *or* Interferences. *The measurement of Weight is made precise by the* Method of Double-weighing. *Secondary Qualities are measured by means of* Scales of Degrees; *but in order to apply these Scales, the student requires the* Education of the Senses. *The Education of the Senses is forwarded by the practical study of* Descriptive Natural History, Chemical Manipulation, *and* Astronomical Observation.

1. I shall speak, in this chapter, of Methods of exact and systematic observation, by which such facts are collected as form the materials of precise scientific propositions. These Methods are very various, according to the nature of the subject inquired into, and other circumstances: but a great portion of them agree in being processes of measurement. These I shall peculiarly consider: and in the first place those referring to Number, Space, and Time, which are at the same time objects and instruments of measurement.

2. But though we have to explain how observations may be made as perfect as possible, we must not forget that in most cases complete perfection is unattainable. *Observations are never perfect.* For we observe phenomena by our senses, and measure their relations in time and space; but our senses and our measures are all, from various causes, inaccurate. If we have to observe the exact place of the moon among the stars, how much of instrumental apparatus is necessary! This apparatus has been improved by many successive generations of astronomers, yet it is still far from being perfect. And the senses of man, as well as his implements, are limited in their exactness. Two different observers do not obtain precisely the same measures of the time and place of a phenomenon; as, for instance, of the moment at which the moon occults a star, and the point of her *limb* at which the occultation takes place. Here, then, is a source of inaccuracy and er-

rour, even in astronomy, where the means of exact observation are incomparably more complete than they are in any other department of human research. In other cases, the task of obtaining accurate measures is far more difficult. If we have to observe the tides of the ocean when rippled with waves, we can see the average level of the water first rise and then fall; but how hard is it to select the exact moment when it is at its greatest height, or the exact highest point which it reaches! It is very easy, in such a case, to err by many minutes in time, and by several inches in space.

Still, in many cases, good Methods can remove very much of this inaccuracy, and to these we now proceed.

3. *(I) Number* Number is the first step of measurement, since it measures itself, and does not, like space and time, require an arbitrary standard. Hence the first exact observations, and the first advances of rigorous knowledge, appear to have been made by means of number; as for example,—the number of days in a month and in a year;—the cycles according to which eclipses occur;—the number of days in the revolutions of the planets; and the like. All these discoveries, as we have seen in the History of Astronomy, go back to the earliest period of the science, anterior to any distinct tradition; and these discoveries presuppose a series, probably a very long series, of observations, made principally by means of number. Nations so rude as to have no other means of exact measurement, have still systems of numeration by which they can reckon to a considerable extent. Very often, such nations have very complex systems, which are capable of expressing numbers of great magnitude. Number supplies the means of measuring other quantities, by the assumption of a *unit* of measure of the appropriate kind: but where nature supplies the unit, number is applicable directly and immediately. Number is an important element in the Classificatory as well as in the Mathematical Sciences. The History of those Sciences shows how the formation of botanical systems was effected by the adoption of number as a leading element, by Cæsalpinus; and how afterwards the Reform of Linnæus in classification depended in a great degree on his finding, in the pistils and stamens, a better numerical basis than those before employed. In like manner, the number of rays in the membrane of the gills,[23] and the number of rays in the fins of fish, were found to be important elements

in ichthyological classification by Artedi and Linnæus. There are innumerable instances, in all parts of Natural History, of the importance of the observation of number. And in this observation, no instrument, scale or standard is needed, or can be applied; except the scale of natural numbers, expressed either in words or in figures, can be considered as an instrument.

4. *(II) Measurement of Space* Of quantities admitting of *continuous* increase and decrease, (for number is discontinuous,) space is the most simple in its mode of measurement, and requires most frequently to be measured. The obvious mode of measuring space is by the repeated application of a material measure, as when we take a foot-rule and measure the length of a room. And in this case the foot-rule is the *unit* of space, and the length of the room is expressed by the number of such units which it contains: or, as it may not contain an exact number, by a number with a *fraction*. But besides this measurement of linear space, there is another kind of space which, for purposes of science, it is still more important to measure, namely, angular space. The visible heavens being considered as a sphere, the portions and paths of the heavenly bodies are determined by drawing circles on the surface of this sphere, and are expressed by means of the parts of these circles thus intercepted: by such measures the doctrines of astronomy were obtained in the very beginning of the science. The arcs of circles thus measured, are not like linear spaces, reckoned by means of an *arbitrary* unit; for there is a *natural unit*, the total circumference, to which all arcs may be referred. For the sake of convenience, the whole circumference is divided into 360 parts or *degrees;* and by means of these degrees and their parts, all arcs are expressed. The *arcs* are the measures of the *angles at the center,* and the degrees may be considered indifferently as measuring the one or the other of these quantities.

5. In the History of Astronomy,[24] I have described the method of observation of celestial angles employed by the Greeks. They determined the lines in which the heavenly bodies were seen, by means either of Shadows, or of Sights; and measured the angles between such lines by arcs or rules properly applied to them. The Armill, Astrolabe, Dioptra, and Parallactic Instrument of the ancients, were some of the instruments thus constructed. Tycho Brahe greatly im-

proved the methods of astronomical observation by giving steadiness to the frame of his instruments, (which were large *quadrants,*) and accuracy to the divisions of the *limb.*[25] But the application of the *telescope* to the astronomical quadrant and the fixation of the center of the field by a *cross* of fine wires placed in the focus, was an immense improvement of the instrument, since it substituted a precise visual ray, pointing to the star, instead of the coarse coincidence of Sights. The accuracy of observation was still further increased by applying to the telescope a *micrometer* which might subdivide the smaller divisions of the arc.

6. By this means, the precision of astronomical observation was made so great, that very minute angular spaces would be measured: and it then became a question whether discrepancies which appeared at first as defects in the theory, might not arise sometimes from a bending or shaking of the instrument, and from the degrees marked on the limb being really somewhat unequal, instead of being rigorously equal. Accordingly, the framing and balancing of the instrument, so as to avoid all possible tremor or flexure, and the exact division of an arc into equal parts, became great objects of those who wished to improve astronomical observations. The observer no longer gazed at the stars from a lofty tower, but placed his telescope on the solid ground, and braced and balanced it with various contrivances. Instead of a quadrant, an entire circle was introduced (by Ramsden); and various processes were invented for the dividing of instruments. Among these we may notice Troughton's method of dividing; in which the visual ray of a microscope was substituted for the points of a pair of compasses, and, by *stepping* round the circle, the partial arcs were made to bear their exact relation to the whole circumference.

7. Astronomy is not the only science which depends on the measurement of angles. Crystallography also requires exact measures of this kind; and the *goniometer,* especially that devised by Wollaston, supplies the means of obtaining such measures. The science of Optics also, in many cases, requires the measurement of angles.

8. In the measurement of linear space, there is no natural standard which offers itself. Most of the common measures appear to be taken from some part of the human body; as a *foot,* a *cubit,* a *fathom;* but

such measures cannot possess any precision, and are altered by convention: thus there were in ancient times many kinds of cubits; and in modern Europe, there are a great number of different standards of the foot, as the Rhenish foot, the Paris foot, the English foot. It is very desirable that, if possible, some permanent standard, founded in nature, should be adopted; for the conventional measures are lost in the course of ages; and thus, dimensions expressed by means of them become unintelligible. Two different natural standards have been employed in modern times: the French have referred their measures of length to the total circumference of a meridian of the earth; a quadrant of this meridian consists of ten million units or *metres*. The English have fixed their linear measure by reference to the length of a pendulum which employs an exact second of time in its small oscillation. Both these methods occasion considerable difficulties in carrying them into effect; and are to be considered mainly as means of recovering the standard if it should ever be lost. For common purposes, some material standard is adopted as authority for the time: for example, the standard which in England possessed legal authority up to the year 1835 was preserved in the House of Parliament; and was lost in the conflagration which destroyed that edifice. The standard of length now generally referred to by men of science in England is that which is in the possession of the Astronomical Society of London.

9. A standard of length being established, the artifices for applying it, and for subdividing it in the most accurate manner, are nearly the same as in the case of measures of arcs: as for instance, the employment of the visual rays of microscopes instead of the legs of compasses and the edges of rules; the use of micrometers for minute measurements; and the like. Many different modes of avoiding errour in such measurements have been devised by various observers, according to the nature of the cases with which they had to deal.[26]

10. *(III) Measurement of Time* The methods of measuring Time are not so obvious as the methods of measuring space; for we cannot apply one portion of time to another, so as to test their equality. We are obliged to begin by assuming some change as the measure of time. Thus the motion of the sun in the sky, or the length and position of the shadows of objects, were the first modes of measuring the parts of the day. But what assurance had men, or what assurance

could they have, that the motion of the sun or of the shadow was uniform? They could have no such assurance, till they had adopted some measure of smaller times; which smaller times, making up larger times by repetition, they took as the standard of uniformity;—for example, an hour-glass, or a clepsydra which answered the same purpose among the ancients. There is no apparent reason why the successive periods measured by the emptying of the hour-glass should be unequal; they are implicitly accepted as equal; and by reference to these the uniformity of the sun's motion may be verified. But the great improvement in the measurement of time was the use of a pendulum for the purpose by Galileo, and the application of this device to clocks by Huyghens in 1656. For the successive oscillations of a pendulum are rigorously equal, and a clock is only a train of machinery employed for the purpose of counting these oscillations. By means of this invention, the measure of time in astronomical observations became as accurate as the measure of space.

11. What is the *natural unit* of time? It was assumed from the first by the Greek astronomers, that the sidereal days, measured by the revolution of a star from any meridian to the same meridian again, are exactly equal; and all improvements in the measure of time tended to confirm this assumption. The sidereal day is therefore the natural standard of time. But the solar day, determined by the diurnal revolution of the sun, although not rigorously invariable, as the sidereal day is, undergoes scarcely any perceptible variation; and since the course of daily occurrences is regulated by the sun, it is far more convenient to seek the basis of our unit of time in *his* motions. Accordingly the solar day (the *mean* solar day) is divided into 24 hours, and these, into minutes and seconds; and this is our scale of time. Of such time, the sidereal day has 23 hours 56 minutes 4.09 seconds. And it is plain that by such a statement the length of the hour is fixed, with reference to a sidereal day. The *standard* of time (and the standard of space in like manner) equally answers its purpose, whether or not it coincides with any *whole number* of units.

12. Since the sidereal day is thus the standard of our measures of time, it becomes desirable to refer to it, constantly and exactly, the instruments by which time is measured, in order that we may secure ourselves against errour. For this purpose, in astronomical observa-

tories, observations are constantly made of the transit of stars across the meridian; the *transit instrument* with which this is done being adjusted with all imaginable regard to accuracy.[27]

13. When exact measures of time are required in other than astronomical observations, the same instruments are still used, namely, clocks and chronometers. In chronometers, the regulating part is an oscillating body; not, as in clocks, a pendulum oscillating by the force of gravity, but a wheel swinging to and fro on its center, in consequence of the vibrations of a slender coil of elastic wire. To divide time into still smaller portions than these vibrations, other artifices are used; some of which will be mentioned under the next head.

14. *(IV) Conversion of Space and Time* Space and time agree in being extended quantities, which are made up and measured by the repetition of homogeneous parts. If a body move uniformly, whether in the way of revolving or otherwise, the *space* which any point describes, is *proportional* to the *time* of its motion; and the space and the time may each be taken as a measure of the other. Hence in such cases, by taking space instead of time, or time instead of space, we may often obtain more convenient and precise measures, than we can by measuring directly the element with which we are concerned.

The most prominent example of such a conversion, is the measurement of the Right Ascension of stars, (that is, their angular distance from a standard meridian[28] on the celestial sphere,) by means of the time employed in their coming to the meridian of the place of observation. Since, as we have already stated, the visible celestial sphere, carrying the fixed stars, revolves with perfect uniformity about the pole; if we observe the stars as they come in succession to a fixed circle passing through the poles, the intervals of time between these observations will be proportional to the angles which the meridian circles passing through these stars make at the poles where they meet; and hence, if we have the means of measuring time with great accuracy, we can, by watching the times of the transits of successive stars across some visible mark in our own meridian, determine the *angular distances* of the meridian circles of all the stars from one another.

Accordingly, now that the pendulum clock affords astronomers the means of determining time exactly, a measurement of the Right

Ascensions of heavenly bodies by means of a clock and a transit instrument, is a part of the regular business of an observatory. If the sidereal clock be so adjusted that it marks the beginning of its scale of time when the first point of Right Ascension is upon the visible meridian of our observatory, the point of the scale at which the clock points when any other star is in our meridian, will truly represent the Right Ascension of the star.

Thus as the motion of the stars is our measure of time, we employ time, conversely, as our measure of the places of the stars. The celestial machine and our terrestrial machines correspond to each other in their movements; and the star steals silently and steadily across our meridian line, just as the pointer of the clock steals past the mark of the hour. We may judge of the scale of this motion by considering that the full moon employs about two minutes of time in sailing across any fixed line seen against the sky, transverse to her path: and all the celestial bodies, carried along by the revolving sphere, travel at the same rate.

15. In this case, up to a certain degree, we render our measures of astronomical angles more exact and convenient by substituting time for space; but when, in the very same kind of observation, we wish to proceed to a greater degree of accuracy, we find that it is best done by substituting space for time. In observing the transit of a star across the meridian, if we have the clock within hearing, we can count the beats of the pendulum by the noise which they make, and tell exactly at which second of time the passage of the star across the visible thread takes place: and thus we measure Right Ascension by means of time. But our perception of time does not allow us to divide a second into ten parts, and to pronounce whether the transit takes place three-tenths, six-tenths, or seven-tenths of a second after the preceding beat of the clock. This, however, can be done by the usual mode of observing the transit of a star. The observer, listening to the beat of his clock, fastens his attention upon the star at each beat, and especially at the one immediately before and the one immediately after the passage of the thread: and by this means he has these two positions and the position of the thread so far present to his intuition at once, that he can judge in what proportion the thread is nearer to one position than the other, and can thus divide the intervening second in

its due proportion. Thus if he observe that at the beginning of the second the star is on one side of the thread, and at the end of the second on the other side; and that the two distances from the thread are as two to three, he knows that the transit took place at two-fifths (or four-tenths) of a second after the former beat. In this way a second of time in astronomical observations may, by a skilful observer, be divided into ten equal parts; although when time is observed as time, a tenth of a second appears almost to escape our senses. From the above explanation, it will be seen that the reason why the subdivision is possible in the way thus described, is this:—that the moment of time thus to be divided is so small, that the eye and the mind can retain, to the end of this moment, the impression of position which it received at the beginning. Though the two positions of the star, and the intermediate thread, are seen successively, they can be contemplated by the mind as if they were seen simultaneously: and thus it is precisely the smallness of this portion of time which enables us to subdivide it by means of space.

16. There is another case, of somewhat a different kind, in which time is employed in measuring space; namely, when space, or the standard of space, is defined by the length of a pendulum oscillating in a given time. We might in this way define any space by the time which a pendulum of such a length would take in oscillating; and thus we might speak, as was observed by those who suggested this device, of five minutes of cloth, or a rope half an hour long. We may observe, however, that in this case, the space is *not proportional* to the time. And we may add, that though we thus appear to avoid the arbitrary standard of space (for as we have seen, the standard of measures of time is a natural one,) we do not do so in fact: for we assume the invariableness of gravity, which really varies (though very slightly,) from place to place.

17. *(V) The Method of Repetition in Measurement* In many cases we can give great additional accuracy to our measurements by repeatedly adding to itself the quantity which we wish to measure. Thus if we wished to ascertain the exact breadth of a thread, it might not be easy to determine whether it was one-ninetieth, or one-ninety-fifth, or one-hundredth part of an inch; but if we find that ninety-six such threads placed side by side occupy exactly an inch, we have the

precise measure of the breadth of the thread. In the same manner, if two clocks are going nearly at the same rate, we may not be able to distinguish the excess of an oscillation of one of the pendulums over an oscillation of the other: but when the two clocks have gone for an hour, one of them may have gained ten seconds upon the other; thus showing that the proportion of their times of oscillation is 3610 to 3600.

In the latter of these instances, we have the principle of repetition truly exemplified, because (as has been justly observed by Sir J. Herschel,) there is then 'a juxtaposition of units without errour,'—'one vibration commences exactly where the last terminates, no part of time being lost or gained in the addition of the units so counted.' In space, this juxtaposition of units without errour cannot be rigorously accomplished, since the units must be added together by material Contact (as in the above case of the threads,) or in some equivalent manner. Yet the principle of repetition has been applied to angular measurement with considerable success in Borda's Repeating Circle. In this instrument, the angle between two objects which we have to observe, is repeated along the graduated limb of the circle by turning the telescope from one object to the other, alternately fastened to the circle by its *clamp*) and loose from it (by unclamping). In this manner the errours of graduation may (theoretically) be entirely got rid of: for if an angle repeated *nine* times be found to go twice round the circle, it must be *exactly* eighty degrees: and where the repetition does not give an exact number of circumferences, it may still be made to subdivide the errour to any required extent.

18. Connected with the principle of repetition, is the *Method of coincidences* or *interferences*. If we have two Scales, on one of which an inch is divided into 10, and on the other into 11 equal parts; and if, these Scales being placed side by side, it appear that the beginning of the latter Scale is between the 2nd and 3rd division of the former, it may not be apparent what fraction added to 2 determines the place of the beginning of the second Scale as measured on the first. But if it appear also that the 3rd division of the second Scale *coincides* with a certain division of the first, (the 5th,) it is certain that 2 and *three-tenths* is the *exact* place of the beginning of the second Scale, measured on the first Scale. The 3rd division of the II Scale will coincide

(or interfere with) a division of the 10 Scale, when the beginning or *zero* of the 11 divisions is three-tenths of a division beyond the preceding line of the 10 Scale; as will be plain on a little consideration. And if we have two Scales of equal units, in which each unit is divided into nearly, but not quite, the same number of equal parts (as 10 and 11, 19 and 20, 29 and 30,) and one sliding on the other, it will always happen that some one or other of the division lines will coincide, or very nearly coincide; and thus the exact position of the beginning of one unit, measured on the other scale, is determined. A sliding scale, thus divided for the purpose of subdividing the units of that on which it slides, is called a *Vernier,* from the name of its inventor.

19. The same principle of Coincidence of Interference is applied to the exact measurement of the length of time occupied in the oscillation of a pendulum. If a detached pendulum, of such a length as to swing in little less than a second, be placed before the seconds' pendulum of a clock, and if the two pendulums begin to move together, the former will gain upon the latter, and in a little while their motions will be quite discordant. But if we go on watching, we shall find them, after a time, to agree again exactly; namely, when the detached pendulum has gained one complete oscillation (back and forwards,) upon the clock pendulum, and again coincides with it in its motion. If this happen after 5 minutes, we know that the times of oscillation of the two pendulums are in the proportion of 300 to 302, and therefore the detached pendulum oscillates in 150/151 of a second. The accuracy which can be obtained in the measure of an oscillation by this means is great; for the clock can be compared (by observing transits of the stars or otherwise) with the natural standard of time, the sidereal day. And the moment of coincidence of the two pendulums may, by proper arrangements, be very exactly determined.

We have hitherto spoken of methods of measuring time and space, but other elements also may be very precisely measured by various means.

20. *(VI) Measurement of Weight* Weight, like space and time, is a quantity made up by addition of parts, and may be measured by similar methods. The principle of repetition is applicable to the measurement of weight; for if two bodies be simultaneously put in the same pan of a balance, and if they balance pieces in the other pan, their weights are exactly added.

There may be difficulties of practical workmanship in carrying into effect the mathematical conditions of a perfect balance; for example, in securing an exact equality of the effective arms of the beam in all positions. These difficulties are evaded by the *Method of double weighing;* according to which the standard weights, and the body which is to be weighed, are successively put in the *same* pan, and made to balance by a third body in the opposite scale. By this means the different lengths of the arms of the beam, and other imperfections of the balance, become of no consequence.

21. There is no natural *Standard* of weight. The conventional weight taken as the standard, is the weight of a given bulk of some known substance; for instance, a *cubic foot of water.* But in order that this may be definite, the water must not contain any portion of heterogenous substance: hence it is required that the water be *distilled* water.

22. *(VII) Measurement of Secondary Qualities* We have already seen[29] that secondary qualities are estimated by means of conventional Scales, which refer them to space, number, or some other definite expression. Thus the Thermometer measures heat; the Musical Scale, with or without the aid of number, expresses the pitch of a note; and we may have an exact and complete Scale of Colours, pure and impure. We may remark, however, that with regard to sound and colour, the estimates of the ear and the eye are not superseded, but only assisted: for it we determine what a note is, by comparing it with an instrument known to be in tune, we still leave the ear to decide when the note is *in unison* with one of the notes of the instrument. And when we compare a colour with our chromatometer, we judge by the eye which division of the chromatometer it *matches.* Colour and sound have their Natural Scales, which the eye and ear habitually apply; what science requires is, that those scales should be systematized. We have seen that several conditions are requisite in such scales of qualities: the observer's skill and ingenuity are mainly shown in devising such scales and methods of applying them.

23. The Method of Coincidences is employed in harmonics: for if two notes are nearly, but not quite, in unison, the coincidences of the vibrations produce an audible undulation in the note, which is called the *howl;* and the exactness of the unison is known by this howl vanishing.

24. *(VIII) Manipulation* The process of applying practically methods of experiment and observation, is termed Manipulation; and the value of observations depends much upon the proficiency of the observer in this art. This skill appears, as we have said, not only in devising means and modes in measuring results, but also in inventing and executing arrangements by which elements are subjected to such conditions as the investigation requires: in finding and using some material combination by which nature shall be asked the question which we have in our minds. To do this in any subject may be considered as a peculiar Art, but especially in Chemistry; where 'many experiments, and even whole trains of research, are essentially dependent for success on mere manipulation.'[30] The changes which the chemist has to study,—compositions, decompositions, and mutual actions, affecting the internal structure rather than the external form and motion of bodies,—are not familiarly recognized by common observers, as those actions are which operate upon the total mass of a body: and hence it is only when the chemist has become, to a certain degree, familiar with his science, that he has the power of observing. He must learn to interpret the effects of mixture, heat, and other Chemical agencies, so as to see in them those facts which chemistry makes the basis of her doctrines. And in learning to interpret this language, he must also learn to call it forth;—to place bodies under the requisite conditions, by the apparatus of his own laboratory and the operations of his own fingers. To do this with readiness and precision, is, as we have said, an Art, both of the mind and of the hand, in no small degree recondite and difficult. A person may be well acquainted with all the doctrines of chemistry, and may yet fail in the simplest experiment. How many precautions and observances, what resource and invention, what delicacy and vigilance, are requisite in *Chemical Manipulation,* may be seen by reference to Dr. Faraday's work on that subject.

25. The same qualities in the observer are requisite in some other departments of science; for example, in the researches of Optics: for in these, after the first broad facts have been noticed, the remaining features of the phenomena are both very complex and very minute; and require both ingenuity in the invention of experiments, and a keen scrutiny of their results. We have instances of the application of

these qualities in most of the optical experimenters of recent times, and certainly in no one more than Sir David Brewster. Omitting here all notice of his succeeding labours, his *Treatise on New Philosophical Instruments,* published in 1813, is an excellent model of the kind of resource and skill of which we now speak. I may mention as an example of this skill, his mode of determining the refractive power of an *irregular* fragment of any transparent substance. At first this might appear an impossible problem; for it would seem that a regular and smooth surface are requisite, in order that we may have any measurable refraction. But Sir David Brewster overcame the difficulty by immersing the fragment in a combination of fluids, so mixed, that they had the same refractive power as the specimen. The question, *when* they had this power, was answered by noticing when the fragment became so transparent that its surface could hardly be seen; for this happened when, the refractive power within and without the fragment being the same, there was no refraction at the surface. And this condition being obtained, the refractive power of the fluid, and therefore of the fragment, was easily ascertained.

26. *(IX) The Education of the Senses* Colour and Musical Tone are, as we have seen, determined by means of the Senses, whether or not Systematical Scales are used in expressing the observed fact. Systematical Scales of sensible qualities, however, not only give precision to the record, but to the observation. But for this purpose such an Education of the Senses is requisite as may enable us to apply the scale immediately. The memory must retain the sensation or perception to which the technical term or degree of the scale refers. Thus with regard to colour, as we have said already, when we find such terms as *tin-white* or *pinchbeck-brown,* the metallic colour so denoted ought to occur at once to our recollection without delay or search. The observer's senses, therefore, must be educated, at first by an actual exhibition of the standard, and afterwards by a familiar use of it, to understand readily and clearly each phrase and degree of the scales which in his observations he has to apply. This is not only the best, but in many cases the only way in which the observation can be expressed. Thus *glassy lustre, fatty lustre, adamantine lustre,* denote certain kinds of shining in minerals, which appearances we should endeavour in vain to describe by periphrasis; and which the terms, if

considered as terms in common language, would by no means clearly discriminate: for who, in common language, would say that coal has a fatty lustre? But these terms, in their conventional sense, are perfectly definite; and when the eye is once familiarized with this application of them, are easily and clearly intelligible.

27. The education of the senses, which is thus requisite in order to understand well the terminology of any science, must be acquired by an inspection of the objects which the science deals with; and is, perhaps, best promoted by the practical study of Natural History. In the different departments of Natural History, the descriptions of species are given by means of an extensive technical *terminology:* and that education of which we now speak, ought to produce the effect of making the observer as familiar with each of the terms of this terminology as we are with the words of our common language. The technical terms have a much more precise meaning that other terms, since they are defined by express convention, and not learnt by common usage merely. Yet though they are thus defined, not the definition, but the perception itself, is that which the term suggests to the proficient.

In order to use the terminology to any good purpose, the student must possess it, not as a dictionary, but as a language. The terminology of his sciences must be the natural historian's most familiar tongue. He must learn to think in such language. And when this is achieved, the terminology, as I have elsewhere said, though to an uneducated eye cumbrous and pedantical, is felt to be a useful implement, not an oppressive burden.[31] The impatient schoolboy looks upon his grammar and vocabulary as irksome and burdensome; but the accomplished student who has learnt the language by means of them, knows that they have given him the means of expressing what he thinks, and even of thinking more precisely. And as the study of language thus gives precision to the thoughts, the study of Natural History, and especially of the descriptive part of it, gives precision to the senses.

The Education of the Senses is also greatly promoted by the practical pursuit of any science of experiment and observation, as chemistry or astronomy. The methods of manipulating, of which we have just spoken, in chemistry, and the methods of measuring extremely minute portions of space and time which are employed in astronomy, and which are described in the former part of this chapter, are among the best modes of educating the senses for purposes of scientific observation.

28. By the various Methods of precise observation which we have thus very briefly described, facts are collected, of an exact and definite kind; they are then bound together in general laws, by the aid of general ideas and of such methods as we have now to consider. It is true, that the ideas which enable us to combine facts into general propositions, do commonly operate in our minds while we are still engaged in the office of observing. Ideas of one kind or other are requisite to connect our phenomena into facts, and to give meaning to the terms of our descriptions: and it frequently happens, that long before we have collected all the facts which induction requires, the mind catches the suggestion which some of these ideas offer, and leaps forwards to a conjectural law while the labour of observation is yet unfinished. But though this actually occurs, it is easy to see that the process of combining and generalizing facts is, in the order of nature, posterior to, and distinct from, the process of observing facts. Not only is this so, but there is an intermediate step which, though inseparable from all successful generalization, may be distinguished from it in our survey; and may, in some degree, be assisted by peculiar methods. To the consideration of such methods we now proceed.

## CHAPTER III
## OF METHODS OF ACQUIRING CLEAR SCIENTIFIC IDEAS; *and first* OF INTELLECTUAL EDUCATION

### *Aphorism XXIX*

*The Methods by which the acquisition of clear Scientific Ideas is promoted, are mainly two;* Intellectual Education *and* Discussion of Ideas.

### *Aphorism XXX*

*The Idea of Space becomes more clear by studying* Geometry; *the Idea of Force, by studying* Mechanics; *the Ideas of Likeness, of Kind, of Subordination of Classes, by studying* Natural History.

*Aphorism XXXI*

Elementary Mechanics *should now form a part of intellectual education, in order that the student may understand the Theory of Universal Gravitation: for an intellectual education should cultivate such ideas as enable the student to understand the most complete and admirable portions of the knowledge which the human race has attained to.*

*Aphorism XXXII*

Natural History *ought to form a part of intellectual education, in order to correct certain prejudices which arise from cultivating the intellect by means of mathematics alone; and in order to lead the student to see that the division of things into Kinds, and the attribution and use of Names, are processes susceptible of great precision. . . .*

## CHAPTER IV
## OF METHODS OF ACQUIRING CLEAR SCIENTIFIC IDEAS, *continued*—OF THE DISCUSSION OF IDEAS

*Aphorism XXXIII*

*The conceptions involved in scientific truths have attained the requisite degree of clearness by means of the* Discussions *respecting ideas which have taken place among discoverers and their followers. Such discussions are very far from being unprofitable to science. They are* metaphysical, *and must be so: the difference between discoverers and barren reasoners is, that the former employ good, and the latter bad metaphysics. . . .*

## CHAPTER V
## ANALYSIS OF THE PROCESS OF INDUCTION

*Aphorism XXXIV*

*The Process of Induction may be resolved into three steps; the* Selection of the Idea, *the* Construction of the Conception, *and the* Determination of the Magnitudes.

*Aphorism XXXV*

*These three steps correspond to the determination of the* Independent Variable, *the* Formula, *and the* Coefficients, *in mathematical investigations; or to the* Argument, *the* Law, *and the* Numerical Data, *in a Table of an astronomical or other* Inequality.

#### *Aphorism* XXXVI

*The Selection of the Idea depends mainly upon inventive sagacity: which operates by suggesting and trying various hypotheses. Some inquirers try erroneous hypotheses; and thus, exhausting the forms of errours, form the Prelude to Discovery.*

#### *Aphorism* XXXVII

*The following Rules may be given, in order to the selection of the Idea for purposes of Induction:—the Idea and the Facts must be* homogeneous; *and the Rule must be* tested by the Facts.

#### *Sect. I. The Three Steps of Induction*

1. When facts have been decomposed and phenomena measured, the philosopher endeavours to combine them into general laws, by the aid of Ideas and Conceptions; these being illustrated and regulated by such means as we have spoken of in the last two chapters. In this task, of gathering laws of nature from observed facts, as we have already said,[32] the natural sagacity of gifted minds is the power by which the greater part of the successful results have been obtained; and this power will probably always be more efficacious than any Method can be. Still there are certain methods of procedure which may, in such investigations, give us no inconsiderable aid, and these I shall endeavour to expound.

2. For this purpose, I remark that the Colligation of ascertained Facts into general Propositions may be considered as containing three steps, which I shall term *the Selection of the Idea, the Construction of the Conception,* and *the Determination of the Magnitudes.* It will be recollected that by the word *Idea,* (or Fundamental Idea,) used in a peculiar sense, I mean certain wide and general fields of intelligible relation, such as Space, Number, Cause, Likeness; while by *Conception* I denote more special modifications of these ideas, as a *circle,* a *square number,* a *uniform force,* a *like form* of flower. Now in order to establish any law by reference to facts, we must select the *true Idea* and the *true Conception.* For example; when Hipparchus found[33] that the distance of the bright star Spica Virginis from the equinoxial point

had increased by two degrees in about two hundred years, and desired to reduce this change to a law, he had first to assign, if possible, the *idea* on which it depended;—whether it was regulated for instance, by *space,* or by *time;* whether it was determined by the positions of other stars at each moment, or went on progressively with the lapse of ages. And when there was found reason to select *time* as the regulative *idea* of this change, it was then to be determined how the change went on with the time;—whether uniformly, or in some other manner: the *conception,* or the rule of the progression, was to be rightly constructed. Finally, it being ascertained that the change did go on uniformly, the question then occurred what was its *amount:*—whether exactly a degree in a century, or more, or less, and how much: and thus the determination of the *magnitude* completed the discovery of the law of phenomena respecting this star.

3. Steps similar to these three may be discerned in all other discoveries of laws of nature. Thus, in investigating the laws of the motions of the sun, moon or planets, we find that these motions may be resolved, besides a uniform motion, into a series of partial motions, or Inequalities; and for each of these Inequalities, we have to learn upon what it directly depends, whether upon the progress of time only, or upon some configuration of the heavenly bodies in space; then, we have to ascertain its law; and finally, we have to determine what is its amount. In the case of such Inequalities, the fundamental element on which the Inequality depends, is called by mathematicians the *Argument.* And when the Inequality has been fully reduced to known rules, and expressed in the form of a Table, the argument is the fundamental Series of Numbers which stands in the margin of the Table, and by means of which we refer to the other Numbers which express the Inequality. Thus, in order to obtain from a Solar Table the Inequality of the sun's annual motion, the Argument is the Number which expresses the day of the year; the Inequalities for each day being (in the Table) ranged in a line corresponding to the days. Moreover, the Argument of an Inequality being assumed to be known, we must, in order to calculate the Table, that is, in order to exhibit the law of nature, know also the *Law* of the Inequality, and its *Amount.* And the investigation of these three things, the Argument, the Law, and the Amount of the Inequality, represents the three steps above described, the Selection

of the Idea, the Construction of the Conception, and the Determination of the Magnitude.

4. In a great body of cases, *mathematical* language and calculation are used to express the connexion between the general law and the special facts. And when this is done, the three steps above described may be spoken of as the Selection of the *Independent Variable,* the Construction of the *Formula,* and the Determination of the *Coefficients.* It may be worth our while to attend to an exemplification of this. Suppose then, that, in such observations as we have just spoken of, namely, the shifting of a star from its place in the heavens by an unknown law, astronomers had, at the end of three successive years, found that the star had removed by 3, by 8, and by 15 minutes from its original place. Suppose it to be ascertained also, by methods of which we shall hereafter treat, that this change depends upon the time; we must then take the *time,* (which we may denote by the symbol $t$,) for the *independent variable.* But though the star changes its place *with* the time, the change is not *proportional* to the time; for its motion which is only 3 minutes in the first year, is 5 minutes in the second year, and 7 in the third. But it is not difficult for a person a little versed in mathematics to perceive that the series 3, 8, 15, may be obtained by means of two terms, one of which is proportional to the time, and the other to the square of the time; that is, it is expressed by the *formula* $at + btt$. The question then occurs, what are the values of the *coefficients* $a$ and $b$; and a little examination of the case shows us that $a$ must be 2, and $b$, 1: so that the formula is $2t + tt$. Indeed if we add together the series 2, 4, 6, which expresses a change proportional to the time, and 1, 4, 9, which is proportional to the square of the time, we obtain the series 3, 8, 15, which is the series of numbers given by observation. And thus the three steps which give us the Idea, the Conception, and the Magnitudes; or the Argument, the Law, and the Amount, of the change; give us the Independent Variable, the Formula, and the Coefficients, respectively.

We now proceed to offer some suggestions of methods by which each of these steps may be in some degree promoted.

***Sect. II. Of the Selection of the Fundamental Idea***

5. When we turn our thoughts upon any assemblage of facts, with

a view of collecting from them some connexion or law, the most important step, and at the same time that in which rules can least aid us, is the Selection of the Idea by which they are to be collected. So long as this idea has not been detected, all seems to be hopeless confusion or insulated facts; when the connecting idea has been caught sight of, we constantly regard the facts with reference to their connexion, and wonder that it should be possible for any one to consider them in any other point of view.

Thus the different seasons, and the various aspects of the heavenly bodies, might at first appear to be direct manifestations from some superior power, which man could not even understand: but it was soon found that the ideas of time and space, of motion and recurrence, would give coherency to many of the phenomena. Yet this took place by successive steps. Eclipses, for a long period, seemed to follow no law; and being very remarkable events, continued to be deemed the indications of a supernatural will, after the common motions of the heavens were seen to be governed by relations of time and space. At length, however, the Chaldeans discovered that, after a period of eighteen years, similar sets of eclipses recur; and, thus selecting the idea of *time,* simply, as that to which these events were to be referred, they were able to reduce them to rule; and from that time, eclipses were recognized as parts of a regular order of things. We may, in the same manner, consider any other course of events, and may enquire by what idea they are bound together. For example, if we take the weather, years peculiarly wet or dry, hot and cold, productive and unproductive, follow each other in a manner which, at first sight at least, seems utterly lawless and irregular. Now can we in any way discover some rule and order in these occurrences? Is there, for example, in these events, as in eclipses, a certain cycle of years, after which like seasons come round again? or does the weather depend upon the force of some extraneous body—for instance, the moon—and follow in some way her aspects? or would the most proper way of investigating this subject be to consider the effect of the moisture and heat of various tracts of the earth's surface upon the ambient air? It is at our choice to *try* these and other modes of obtaining a science of the weather: that is, we may refer the phenomena to the idea of *time,* introducing the conception of the moon's action;—or to the idea of *mutual action,* introducing the conceptions of thermotical and atmological agencies, operating between different regions of earth, water, and air.

6. It may be asked, How are we to decide in such alternatives? How are we to select the one right idea out of several conceivable ones? To which we can only reply, that this must be done by *trying* which will succeed. If there really exist a cycle of the weather, as well as of eclipses, this must be established by comparing the asserted cycle with a good register of the seasons, of sufficient extent. Or if the moon really influence the meteorological conditions of the air, the asserted influence must be compared with the observed facts, and so accepted or rejected. When Hipparchus had observed the increase of longitude of the stars, the idea of a motion of the celestial sphere suggested itself as the explanation of the change; but this thought was *verified* only by observing several stars. It was conceivable that each star should have an independent motion, governed by time only, or by other circumstances, instead of being regulated by its place in the sphere; and this possibility could be rejected by trial alone. In like manner, the original opinion of the composition of bodies supposed the compounds to derive their properties from the elements according to the law of *likeness;* but this opinion was overturned by a thousand facts; and thus the really applicable Idea of Chemical Composition was introduced in modern times. In what has already been said on the History of Ideas, we have seen how each science was in a state of confusion and darkness till the right idea was introduced.

7. No general method of evolving such ideas can be given. Such events appear to result from a peculiar sagacity and felicity of mind;—never without labour, never without preparation;—yet with no constant dependence upon preparation, or upon labour, or even entirely upon personal endowments. Newton explained the colours which refraction produces, be referring each colour to a peculiar *angle of refraction,* thus introducing the right idea. But when the same philosopher tried to explain the colours produced by diffraction, he erred, by attempting to apply the same idea, *(the course of a single ray,)* Instead of applying the truer idea, of the *interference of two rays.* Newton gave a wrong rule for the double refraction of Iceland spar, by making the refraction depend on the *edges* of the rhombohedron: Huyghens, more happy, introduced the idea of the *axis of symmetry* of the solid, and thus was able to give the true law of the phenomena.

8. Although the selected idea is proved to be the right one, only

when the true law of nature is established by means of it, yet it often happens that there prevails a settled conviction respecting the relation which must afford the key to the phenomena, before the selection has been confirmed by the laws to which it leads. Even before the empirical laws of the tides were made out, it was not doubtful that these laws depended upon the places and motions of the sun and moon. We know that the crystalline form of a body must depend upon its chemical composition, though we are as yet unable to assign the law of this dependence.

Indeed in most cases of great discoveries, the right idea to which the facts were to be referred, was selected by many philosophers before the decisive demonstration that it *was* the right idea, was given by the discoverer. Thus Newton showed that the motions of the planets might be explained by means of a central force in the sun: but though he established, he did not first select the idea involved in the conception of a central force. The idea had already been sufficiently pointed out, dimly by Kepler, more clearly by Borelli, Huyghens, Wren, and Hooke. Indeed this anticipation of the true idea is always a principal part of that which, in the *History of the Sciences,* we have termed the *Prelude* of a Discovery. The two steps of *proposing* a philosophical problem, and of *solving* it, are, as we have elsewhere said, both important, and are often performed by different persons. The former step is, in fact, the Selection of the Idea. In explaining any change, we have to discover first the *Argument,* and then the *Law* of the change. The selection of the Argument is the step of which we here speak; and is that in which inventiveness of mind and justness of thought are mainly shown.

9. Although, as we have said, we can give few precise directions for this cardinal process, the Selection of the Idea, in speculating on phenomena, yet there is one Rule which may have its use: it is this:— *The idea and the facts must be homogeneous:* the elementary Conceptions, into which the facts have been decomposed, must be of the same nature as the Idea by which we attempt to collect them into laws. Thus, if facts have been observed and measured by reference to space, they must be bound together by the idea of space: if we would obtain a knowledge of mechanical forces in the solar system, we must observe mechanical phenomena. Kepler erred against this rule in his attempts at obtaining physical laws of the system; for the facts which he

took were the *velocities,* not *the changes of velocity,* which are really the mechanical facts. Again, there has been a transgression of this Rule committed by all chemical philosophers who have attempted to assign the relative position of the elementary particles of bodies in their component molecules. For their purpose has been to discover the *relations* of the particles *in space;* and yet they have neglected the only facts in the constitution of bodies which have a reference to space—namely, *crystalline form,* and *optical properties.* No progress can be made in the theory of the elementary structure of bodies, without making these classes of facts the main basis of our speculations.

10. The only other Rule which I have to offer on this subject, is that which I have already given:—*the Idea must be tested by the facts.* It must be tried by applying to the facts the conceptions which are derived from the idea, and not accepted till some of these succeed in giving the law of the phenomena. The justice of the suggestion cannot be known otherwise than by making the trial. If we can discover a *true law* by employing any conceptions, the idea from which these conceptions are derived is the *right* one; nor can there be any proof of its rightness so complete and satisfactory, as that we are by it led to a solid and permanent truth.

This, however, can hardly be termed a Rule; for when we would know, to conjecture and to try the truth of our conjecture by a comparison with the facts, is the natural and obvious dictate of common sense.

Supposing the Idea which we adopt, or which we would try, to be now fixed upon, we still have before us the range of many Conceptions derived from it; many Formulæ may be devised depending on the same Independent Variable, and we must now consider how our selection among these is to be made.

## CHAPTER VI
## GENERAL RULES FOR THE CONSTRUCTION OF THE CONCEPTION

*Aphorism XXXVIII*

*The Construction of the Conception very often includes, in a great measure, the Determination of the Magnitudes.*

***Aphorism XXXIX***

*When a series of* progressive *numbers is given as the results of observation, it may generally be reduced to law by combinations of arithmetical and geometrical progressions.*

***Aphorism XL***

*A true formula for a progressive series of numbers cannot commonly be obtained from a* narrow range *of observations.*

***Aphorism XLI***

Recurrent *series of numbers must, in most cases, be expressed by circular formulæ.*

***Aphorism XLII***

*The true construction of the conception is frequently suggested by some hypothesis; and in these cases, the hypothesis may be useful, though containing superfluous parts.*

1. In speaking of the discovery of laws of nature, those which depend upon *quantity,* as number, space, and the like, are most prominent and most easily conceived, and therefore in speaking of such researches, we shall often use language which applies peculiarly to the cases in which quantities numerically measurable are concerned, leaving it for a subsequent task to extend our principles to ideas of other kinds.

Hence we may at present consider the Construction of a Conception which shall include and connect the facts, as being the construction of a Mathematical Formula, coinciding with the numerical expression of the facts; and we have to consider how this process can be facilitated, it being supposed that we have already before us the numerical measures given by observation.

2. We may remark, however, that the construction of the right Formula for any such case, and the determination of the Coefficients of such formula, which we have spoken of as two separate steps, are in practice almost necessarily simultaneous; for the near coincidence of the results of the theoretical rule with the observed facts confirms at the same time the Formula and its Coefficients. In this case also, the mode of arriving at truth is to try various hypotheses;—to modify the hypotheses so as to approximate to the facts, and to multiply the facts

so as to test the hypotheses.

The Independent Variable, and the Formula which we would try, being once selected, mathematicians have devised certain special and technical processes by which the value of the coefficients may be determined. These we shall treat of in the next Chapter; but in the mean time we may note, in a more general manner, the mode in which, in physical researches, the proper formula may be obtained.

3. A person somewhat versed in mathematics, having before him a series of numbers, will generally be able to devise a formula which approaches near to those numbers. If, for instance, the series is constantly progressive, he will be able to see whether it more nearly resembles an arithmetical or a geometrical progression. For example, MM. Dulong and Petit, in their investigation of the law of cooling of bodies, obtained the following series of measures. A thermometer, made hot, was placed in an enclosure of which the temperature was 0 degrees, and the rapidity of cooling of the thermometer was noted for many temperatures. It was found that

| | | | |
|---|---|---|---|
| For the temperature | 240 | the rapidity of cooling was | 10·69 |
| " | 220 | " | 8·81 |
| " | 200 | " | 7·40 |
| " | 180 | " | 6·10 |
| " | 160 | " | 4·89 |
| " | 140 | " | 3·88 |

and so on. Now this series of numbers manifestly increases with greater rapidity as we proceed from the lower to the higher parts of the scale. The numbers do not, however, form a geometrical series, as we may easily ascertain. But if we were to take the differences of the successive terms we should find them to be—

1·88, 1·41, 1·30, 1·21, 1·01, &c.

and these numbers are very nearly the terms of a geometric series. For if we divide each term by the succeeding one, we find these numbers,

1·33, 1·09, 1·07, 1·20, 1·27,

in which there does not appear to be any constant tendency to diminish or increase. And we shall find that a geometrical series in which the ratio is 1·165, may be made to approach very near to this series, the deviations from it being only such as may be accounted for by conceiving them as errours of observation. In this manner a certain formula[34] is obtained, giving results which very nearly coincide with the observed facts, as may be seen in the margin.

The physical law expressed by the formula just spoken of is this:—that when a body is cooling in an empty inclosure which is kept at a constant temperature, the quickness of the cooling, for excesses of temperature in arithmetical progression, increases as the terms of a geometrical progression, diminished by a constant number.

4. In the actual investigation of Dulong and Petit, however, the formula was not obtained in precisely the manner just described. For the quickness of cooling depends upon two elements, the temperature of the hot body and the temperature of the inclosure; not merely upon the *excess* of one of these over the other. And it was found most convenient, first, to make such experiments as should exhibit the dependence of the velocity of cooling upon the temperature of the enclosure; which dependence is contained in the following law:—The quickness of cooling of a thermometer in vacuo for a constant excess of temperature, increases in geometric progression, when the temperature of the inclosure increases in arithmetic progression. From this law the preceding one follows by necessary consequence.[35]

This example may serve to show the nature of the artifices which may be used for the construction of formulæ, when we have a constantly progressive series of numbers to represent. We must not only endeavour by trial to contrive a formula which will answer the conditions, but we must vary our experiments so as to determine, first one factor or portion of the formula, and then the other; and we must use the most probable hypothesis as means of suggestion for our formulæ.

5. In a *progressive* series of numbers, unless the formula which we adopt be really that which expresses the law of nature, the deviations of the formula from the facts will generally become enormous, when the experiments are extended into new parts of the scale. True formulæ for a progressive series of results can hardly ever be obtained from a very limited range of experiments: just as the attempt to guess the general course of a road or a river, by knowing two or three points of it

in the neighbourhood of one another, would generally fail. In the investigation respecting the laws of the cooling of bodies just noticed, one great advantage of the course pursued by the experimenters was, that their experiments included so great a range of temperatures. The attempts to assign the law of elasticity of steam deduced from experiments made with moderate temperatures, were found to be enormously wrong, when very high temperatures were made the subject of experiment. It is easy to see that this must be so: an arithmetical and a geometrical series may nearly coincide for a few terms moderately near each other: but if we take remote corresponding terms in the two series, one of these will be very many times the other. And hence, from a narrow range of experiments, we may infer one of these series when we ought to infer the other; and thus obtain a law which is widely erroneous.

6. In Astronomy, the serieses of observations which we have to study are, for the most part, not progressive, but *recurrent*. The numbers observed do not go on constantly increasing; but after increasing up to a certain amount they diminish; then, after a certain space, increase again; and so on, changing constantly through certain *cycles*. In cases in which the observed numbers are of this kind, the formula which expresses them must be a *circular function*, of some sort or other; involving, for instance, sines, tangents, and other forms of calculation, which have recurring values when the angle on which they depend goes on constantly increasing. The main business of formal astronomy consists in resolving the celestial phenomena into a series of *terms* of this kind, in detecting their *arguments*, and in determining their *coefficients*.

7. In constructing the formulæ by which laws of nature are expressed, although the first object is to assign the Law of the Phenomena, philosophers have, in almost all cases, not proceeded in a purely empirical manner, to connect the observed numbers by some expression of calculation, but have been guided, in the selection of their formula, by some *Hypothesis* respecting the mode of connexion of the facts. Thus the formula of Dulong and Petit above given was suggested by the Theory of Exchanges; the first attempts at the resolution of the heavenly motions into circular functions were clothed in

the hypothesis of Epicycles. And this was almost inevitable. 'We must confess,' says Copernicus,[36] 'that the celestial motions are circular, or compounded of several circles, since their inequalities observe a fixed law, and recur in value at certain intervals, which could not be except they were circular: for a circle alone can make that quantity which has occurred recur again.' In like manner the first publication of the *Law of the Sines,* the true formula of optical refraction, was accompanied by Descartes with an hypothesis, in which an explanation of the law was pretended. In such cases, the mere comparison of observations may long fail in suggesting the true formulæ. The Fringes of shadows and other diffracted colours were studied in vain by Newton, Gramaldi, Comparetti, the elder Herschel, and Mr. Brougham, so long as these inquirers attempted merely to trace the laws of the facts as they appeared in themselves; while Young, Fresnel, Fraunhofer, Schwerdt, and others, determined these laws in the most rigorous manner, when they applied to the observations the Hypothesis of Interferences.

8. But with all the aid that Hypotheses and Calculation can afford, the construction of true formulæ, in those cardinal discoveries by which the progress of science has mainly been caused, has been a matter of great labour and difficulty, and of good fortune added to sagacity. In the *History of Science,* we have seen how long and how hard Kepler laboured, before he converted the formula for the planetary motions, from an *epicyclical* combination, to a simple *ellipse.* The same philosopher, labouring with equal zeal and perseverance to discover the formula of optical refraction, which now appears to us so simple, was utterly foiled. Malus sought in vain the formula determining the Angle at which a transparent surface polarizes light: Sir D. Brewster,[37] with a happy sagacity, discovered the formula to be simply this, that the *index* of refraction is the *tangent* of the angle of polarization.

Though we cannot give rules which will be of much service when we have thus to divine the general form of the relation by which phenomena are connected, there are certain methods by which, in a narrower field, our investigations may be materially promoted;—certain special methods of obtaining Laws from Observations. Of these we shall now proceed to treat.

## CHAPTER VII
## SPECIAL METHODS OF INDUCTION APPLICABLE TO QUANTITY

### *Aphorism XLIII*

*There are special Methods of Induction applicable to Quantity; of which the principal are, the* Method of Curves, *the* Method of Means, *the* Method of Least Squares, *and the* Method of Residues.

### *Aphorism XLIV*

The Method of Curves *consists in drawing a curve, of which the observed quantities are the Ordinates, the quantity on which the change of these quantities depends being the Abscissa. The efficacy of this Method depends upon the faculty which the eye possesses, of readily detecting regularity and irregularity in forms. The Method may be used to detect the Laws which the observed quantities follow; and also, when the Observations are inexact, it may be used to correct these Observations, so as to obtain data more true than the observed facts themselves.*

### *Aphorism XLV*

The Method of Means *gets rid of irregularities by taking the arithmetical mean of a great number of observed quantities. Its efficacy depends upon this; that in cases in which observed quantities are affected by other inequalities, besides that of which we wish to determine the law, the excesses* above *and defects* below *the quantities which the law in question would produce, will, in a collection of* many *observations,* balance *each other.*

### *Aphorism XLVI*

The Method of Least Squares *is a Method of Means, in which the mean is taken according to the condition, that the sum of the squares of the errours of observation shall be the least possible which the law of the facts allows. It appears, by the Doctrine of changes, that this is the* most probable *mean.*

### *Aphorism XLVII*

The Method of Residues *consists in subtracting, from the quantities given by Observation, the quantity given by any Law already dis-*

*covered; and then examining the remainder, or* Residue, *in order to discover the leading Law which it follows. When this second Law has been discovered, the quantity given by it may be subtracted from the first Residue; thus giving a* Second Residue, *which may be examined in the same manner; and so on. The efficacy of this method depends principally upon the circumstance of the Laws of variation being successively smaller and smaller in amount (or at least in their mean effect); so that the ulterior undiscovered Laws do not prevent the Law in question from being* prominent *in the observations. . .*

***Aphorism XLVIII***

*The Method of Means and the Method of Least Squares cannot be applied without our* knowing the Arguments *of the Inequalities which we seek. The Method of Curves and the Method of Residues, when the Arguments of the principal Inequalities are known, often make it easy to find the others.*

In cases where the phenomena admit of numerical measurement and expression, certain mathematical methods may be employed to facilitate and give accuracy to the determination of the formula by which the observations are connected into laws. Among the most usual and important of these Methods are the following:—

I. The Method of Curves.
II. The Method of Means.
III. The Method of Least Squares.
IV. The Method of Residues.

***Sect. I. The Method of Curves***

1. The Method of Curves proceeds upon this basis; that when one quantity undergoes a series of changes depending on the progress of another quantity, (as, for instance, the Deviation of the Moon from her equable place depends upon the progress of Time,) this dependence may be expressed by means of a *curve*. In the language of mathematicians, the variable quantity, whose changes we would consider, is made the *ordinate* of the curve, and the quantity on which the changes depend is made the *abscissa*. In this manner, the curve will exhibit in its form a series of undulations, rising and falling so as to

correspond with the alternate Increase and Diminution of the quantity represented, at intervals of Space which correspond to the intervals of Time, or other quantity by which the changes are regulated. Thus, to take another example, if we set up, at equal intervals, a series of ordinates representing the Height of all the successive High Waters brought by the tides at a given place, for a year, the curve which connects the summits of all these ordinates will exhibit a series of undulations, ascending and descending once in about each Fortnight; since, in that interval, we have, in succession, the high spring tides and the low neap tides. The curve thus drawn offers to the eye a picture of the order and magnitude of the changes to which the quantity under contemplation, (the height of high water,) is subject.

2. Now the peculiar facility and efficacy of the Method of Curves depends upon this circumstance;—that order and regularity are more readily and clearly recognized, when thus exhibited to the eye in a picture, than they are when presented to the mind in any other manner. To detect the relations of Number considered directly as Number, is not easy: and we might contemplate for a long time a Table of recorded Numbers without perceiving the order of their increase and diminution, even if the law were moderately simple; as any one may satisfy himself by looking at a Tide Table. But if these Numbers are expressed by the magnitude of *Lines,* and if these Lines are arranged in regular order, the eye readily discovers the rule of their changes: it follows the curve which runs along their extremities, and takes note of the order in which its convexities and concavities succeed each other, if any order be readily discoverable. The separate observations are in this manner compared and generalized and reduced to rule by the eye alone. And the eye, so employed, detects relations of order and succession with a peculiar celerity and evidence. If, for example, we thus arrange as ordinates the prices of corn in each year for a series of years, we shall see the order, rapidity, and amount of the increase and decrease of price, far more clearly than in any other manner. And if there were any recurrence of increase and decrease at stated intervals of years, we should in this manner perceive it. The eye, constantly active and busy, and employed in making into shapes the hints and traces of form which it contemplates, runs along the curve thus offered to it; and as it travels backwards and forwards, is ever on the watch to

detect some resemblance or contrast between one part and another. And these resemblances and contrasts, when discovered, are the images of Laws of Phenomena; which are made manifest at once by this artifice, although the mind could not easily catch the indications of their existence, if they were not thus reflected to her in the clear mirror of Space.

Thus when we have a series of good Observations, and know the argument upon which their change of magnitude depends, the Method of Curves enables us to ascertain, almost at a glance, the law of the change; and by further attention, may be made to give us a formula with great accuracy. The Method enables us to perceive, among our observations, an order, which without the method, is concealed in obscurity and perplexity.

3. But the Method of Curves not only enables us to obtain laws of nature from *good* Observations, but also, in a great degree, from observations which are very *imperfect*. For the imperfection of observations may in part be corrected by this consideration;—that though they may appear irregular, the correct facts which they imperfectly represent, are really regular. And the Method of Curves enables us to remedy this apparent irregularity, at least in part. For when Observations thus imperfect are laid down as Ordinates, and their extremities connected by a line, we obtain, not a smooth and flowing curve, such as we should have if the observations contained only the rigorous results of regular laws; but a broken and irregular line, full of sudden and capricious twistings, and bearing on its face marks of irregularities dependent, not upon law, but upon chance. Yet these irregular and abrupt deviations in the curve are, in most cases, but small in extent, when compared with those bendings which denote the effects of regular law. And this circumstance is one of the great grounds of advantage in the Method of Curves. For when the observations thus laid down present to the eye such a broken and irregular line, we can still see, often with great ease and certainty, what twistings of the line are probably due to the irregular errours of observation; and can at once reject these, by drawing a more regular curve, cutting off all such small and irregular sinuosities, leaving some to the right and some to the left; and then proceeding as if this regular curve, and not the irregular one, expressed the observations. In this manner, we suppose the er-

rours of observation to balance each other; some of our corrected measures being too great and others too small, but with no great preponderance either way. We draw our main regular curve, not *through* the points given by our observations, but *among* them: drawing it, as has been said by one of the philosophers[38] who first systematically used this method, 'with a bold but careful hand.' The regular curve which we thus obtain, thus freed from the casual errours of observation, is that in which we endeavour to discover the laws of change and succession.

4. By this method, thus getting rid at once, in a great measure, of errours of observation, we obtain data which are *more true than the* individual *facts themselves.* The philosopher's business is to compare his hypotheses with facts, as we have often said. But if we make the comparison with separate special facts, we are liable to be perplexed or misled, to an unknown amount, by the errours of observation; which may cause the hypothetical and the observed result to agree, or to disagree, when otherwise they would not do so. If, however, we thus take the *whole mass of the facts,* and remove the errours of actual observation,[39] by making the curve which expresses the supposed observation regular and smooth, we have the separate facts corrected by their general tendency. We are put in possession, as we have said, of something more true than any fact by itself is.

One of the most admirable examples of the use of this Method of Curves is found in Sir John Herschel's *Investigation of the Orbits of Double Stars.*[40] The author there shows how far inferior the direct observations of the angle of position are, to the observations corrected by a curve in the manner above stated. 'This curve once drawn,' he says, 'must represent, it is evident, the law of variation of the angle of position, with the time, not only for instants intermediate between the dates of observations, but even at the moments of observation themselves, much better than the individual *raw* observations can possibly (on an average) do. It is only requisite to try a case or two, to be satisfied that by substituting the curve for the points, we have made a nearer approach to nature, and in a great measure eliminated errours of observation.' 'In following the graphical process,' he adds, 'we have a conviction almost approaching to moral certainty that we cannot be greatly misled.' Again, having thus corrected the raw observations, he makes another use of the graphical method, by trying whether an ellipse can be drawn 'if not *through*, at least *among* the points, so as to approach tolerably near them all; and thus approaching to the orbit which is the subject of investigation.'

5. The *Obstacles* which principally impede the application of the Method of Curves are (I) our *ignorance of the argument* of the changes, and (II) the *complication of several laws* with one another.

(I) If we do not know on what quantity those changes depend which we are studying, we may fail entirely in detecting the law of the changes, although we throw the observations into curves. For the true *argument* of the change should, in fact, be made the *abscissa* of the curve. If we were to express, by a series of ordinates, the *hour* of high water on successive days, we should not obtain, or should obtain very imperfectly, the law which these times follow; for the real argument of this change is not the *solar hour,* but the *hour* at which the *moon* passes the meridian. But if we are supposed to be aware that *this* is the *argument,* (which theory suggests and trial instantly confirms) we then do immediately obtain the primary Rules of the Time of High Water, by throwing a series of observations into a Curve, with the Hour of the Moon's Transit for the abscissa.

In like manner, when we have obtained the first great or Semimensual Inequality of the tides, if we endeavour to discover the laws of other Inequalities by means of curves, we must take from theory the suggestion that the Arguments of such inequalities will probably be the *parallax* and the *declination* of the moon. This suggestion again is confirmed by trial; but if we were supposed to be entirely ignorant of the dependence of the changes of the tide on the Distance and Declination of the moon, the curves would exhibit unintelligible and seemingly capricious changes. For by the effect of the Inequality arising from the Paralax, the convexities of the curves which belong to the spring tides, are in some years made alternately greater and less all the year through; while in other years they are made all nearly equal. This difference does not betray its origin, till we refer it to the Paralax; and the same difficulty in proceeding would arise if we were ignorant that the moon's Declination is one of the Arguments of tidal changes.

In like manner, if we try to reduce to law any meteorological changes, those of the Height of the Barometer for instance, we find that we can make little progress in the investigation, precisely because we do not know the Argument on which these changes depend. That there is a certain regular *diurnal* change of small amount, we know; but when we have abstracted this Inequality, (of which the Argument is the *time of day,*) we find far greater Changes left behind, from day to day and from hour to hour; and we express these in curves, but we cannot reduce them to Rule, because we cannot discover on what numerical quantity they depend. The assiduous study of barometrical observations, thrown into curves, may perhaps hereafter point out to us what are the relations of time and space by which these variations are determined; but in the mean time, this subject exemplifies to us our remark, that the method of curves is of comparatively small use, so long as we are in ignorance of the real Arguments of the Inequalities.

6. (II) In the next place, I remark that a difficulty is thrown in the way of the Method of Curves by *the Combination of several laws* one with another. It will readily be seen that such a cause will produce a complexity in the curves which exhibit the succession of facts. If, for example, we take the case of the Tides, the Height of high water increases and diminishes with the Approach of the sun to, and its Recess from, the syzygies of the moon. Again, this Height increases and diminishes as the moon's Parallax increases and diminishes; and again, the Height diminishes when the Declination increases, and *vice versa;* and all these Arguments of change, the Distance from Syzygy, the Parallax, the Declination, complete their circuit and return into themselves in different periods. Hence the curve which represents the Height of high water has not any periodical interval in which it completes its changes and commences a new cycle. The sinuosity which would arise from each Inequality separately considered, interferes with, disguises, and conceals the others; and when we first cast our eyes on the curve of observation, it is very far from offering any obvious regularity in its form. And it is to be observed that we have not yet enumerated *all* the elements of this complexity: for there are changes of the tide depending upon the Parallax and Declination of the Sun as well as of the Moon. Again; besides these changes, of which the Arguments are obvious, there are others, as those depending upon the Barometer and the Wind, which follow no known regular law, and which constantly affect and disturb the results produced by other laws.

In the Tides, and in like manner in the motions of the Moon, we have very eminent examples of the way in which the discovery of laws may be rendered difficult by the number of laws which operate to affect the same quantity. In such cases, the Inequalities are generally picked out in succession, nearly in the order of their magnitudes. In this way there were successively collected, from the study of the Moon's motions by a series of astronomers, those Inequalities which we term the *Equation of the Center*, the *Evection*, the *Variation*, and the *Annual Equation*. These Inequalities were not, in fact, obtained by the application of the Method of Curves; but the Method of Curves might have been applied to such a case with great advantage. The Method has been applied with great industry and with remarkable success to the investigation of the laws of the Tides; and by the use of it, a series of Inequalities both of the Times and of the Heights of high water has been detected, which explain all the main features of the observed facts.

***Sect. II. The Method of Means***

7. The Method of Curves, as we have endeavoured to explain above, frees us from the casual and extraneous irregularities which arise from the imperfection of observation; and thus lays bare the results of the laws which really operate, and enables us to proceed in search of those laws. But the Method of Curves is not the only one which effects such a purpose. The errours arising from detached observations may be got rid of, and the additional accuracy which multiplied observations give may be obtained, by operations upon the observed numbers, without expressing them by spaces. The process of curves assumes that the errours of observation balance each other;—that the accidental excesses and defects are nearly equal in amount;—that the true quantities which would have been observed if all accidental causes of irregularity were removed, are obtained, exactly or nearly, by selecting quantities, upon the whole, equally distant from the extremes of great and small, which our imperfect observations offer to us. But when, among a number of unequal quantities, we take a quantity equally distant from the greater and the smaller, this quantity is termed the *Mean* of the unequal quantities. Hence the correction of our observations by the method of curves consists in taking of Mean of the observations.

8. Now without employing curves, we may proceed arithmetically to take the Mean of all the observed numbers of each class. Thus, if we wished to know the Height of the spring tide at a given place, and if we found that four different spring tides were measured as being of the height of ten, thirteen, eleven, and fourteen feet, we should conclude that the true height of the tide was the *Mean* of these numbers, —namely, twelve feet; and we should suppose that the deviation from this height, in the individual cases, arose from the accidents of weather, the imperfections of observation, or the operation of other laws, besides the alternation of spring and neap tides.

This process of finding the Mean of an assemblage of observed numbers is much practised in discovering, and still more in confirming and correcting, laws of phenomena. We shall notice a few of its peculiarities.

9. The Method of Means requires a knowledge of the *Argument* of the changes which we would study; for the numbers must be arranged in certain Classes, before we find the Mean of each Class; and the principle on which this arrangement depends is the Argument. This knowledge of the Argument is more indispensably necessary in the Method of Means than in the Method of Curves; for when Curves are drawn, the eye often spontaneously detects the law of recurrence in their sinuosities; but when we have collections of Numbers, we must divide them into classes by a selection of our own. Thus, in order to discover the law which the heights of the tide follow, in the progress from spring to neap, we arrange the observed tides according to the *day of the moon's age;* and we then take the mean of all those which thus happen at the *same period* of the Moon's Revolution. In this manner we obtain the law which we seek; and the process is very nearly the same in all other applications of this Method of Means. In all cases, we begin by assuming the Classes of measures which we wish to compare, the Law which we could confirm or correct, the Formula of which we would determine the coefficients.

10. The Argument being thus assumed, the Method of Means is very efficacious in ridding our inquiry of errours and irregularities which would impede and perplex it. Irregularities which are altogether accidental, or at least accidental with reference to some law which we have under consideration, compensate each other in a very remarkable way, when we take the Means of *many* observations. If

we have before us a collection of observed tides, some of them may be elevated, some depressed by the wind, some noted too high and some too low by the observer, some augmented and some diminished by uncontemplated changes in the moon's distance or motion: but in the course of a year or two at the longest, all these causes of irregularity balance each other; and the law of succession, which runs through the observations, comes out as precisely as if those disturbing influences did not exist. In any particular case, there appears to be no possible reason why the deviation should be in one way, or of one moderate amount, rather than another. But taking the mass of observations together, the deviations in opposite ways will be of equal amount, with a degree of exactness very striking. This is found to be the case in all inquiries where we have to deal with observed numbers upon a large scale. In the progress of the population of a country, for instance, what can appear more inconstant, in detail, than the causes which produce births and deaths? yet in each country, and even in each province of a country, the proportions of the whole numbers of births and deaths remain nearly constant. What can be more seemingly beyond the reach of rule than the occasions which produce letters that cannot find their destination? yet it appears that the number of 'dead letters' is nearly the same from year to year. And the same is the result when the deviations arise, not from mere accident, but from laws perfectly regular, though not contemplated in our investigation.[41] Thus the effects of the Moon's Parallax upon the Tides, sometimes operating one way and sometimes another, according to certain rules, are quite eliminated by taking the Means of a long series of observations; the excesses and defects neutralizing each other, so far as concerns the effects upon any law of the tides which we would investigate.

11. In order to obtain very great accuracy, very large masses of observations are often employed by philosophers, and the accuracy of the result increases with the multitude of observations. The immense collections of astronomical observations which have in this manner been employed in order to form and correct the Tables of the celestial motions are perhaps the most signal instances of the attempts to obtain accuracy by this accumulation of observations. Delambre's Tables of the Sun are founded upon nearly 3000 observations; Burg's Tables of the Moon upon above 4000.

But there are other instances hardly less remarkable. Mr. Lubbock's first investigations of the laws of the tides of London,[42] included above 13,000 observations, extending through nineteen years; it being considered that this large number was necessary to remove the effects of accidental causes.[43] And the attempts to discover the laws of change in the barometer have led to the performance of labours of equal amount: Laplace and Bouvard examined this question by means of observations made at the Observatory of Paris, four times every day for eight years.

12. We may remark one striking evidence of the accuracy thus obtained by employing large masses of observations. In this way we may often detect inequalities much smaller than the errours by which they are encumbered and concealed. Thus the Diurnal Oscillations of the Barometer were discovered by the comparison of observations of many days, classified according to the hours of the day; and the result was a clear and incontestable proof of the existence of such oscillations, although the differences which these oscillations produce at different hours of the day are far smaller than the casual changes, hitherto reduced to no law, which go on from hour to hour and from day to day. The effect of law, operating incessantly and steadily, makes itself more and more felt as we give it a longer range; while the effect of accident, followed out in the same manner, is to annihilate itself, and to disappear altogether from the result.

***Sect. III. The Method of Least Squares***

13. The Method of Least Squares is in fact a method of means, but with some peculiar characters. Its object is to determine the *best Mean* of a number of observed quantities; or the *most probable Law* derived from a number of observations, of which some, or all, are allowed to be more or less imperfect. And the method proceeds upon this supposition;—that all errours are not *equally* probable, but that small errours are more probable than large ones. By reasoning mathematically upon this ground, we find that the best result is obtained (since we cannot obtain a result in which the errours vanish) by making, not the *Errours* themselves, but the *Sum of their Squares* of the *smallest* possible amount.

14. An example may illustrate this. Let a quantity which is known

to increase uniformly, (as the distance of a star from the meridian at successive instants,) be measured at equal intervals of time, and be found to be successively 4, 12, 14. It is plain, upon the face of these observations, that they are erroneous; for they ought to form an arithmetical progression, but they deviate widely from such a progression. But the question then occurs, what arithmetical progression do they *most probably* represent: for we may assume several arithmetical progressions which more or less approach the observed series; as for instance, these three; 4, 9, 14; 6, 10, 14; 5, 10, 15. Now in order to see the claims of each of these to the truth, we may tabulate them thus.

| Observation | | | | |
|---|---|---|---|---|
| | 4, 12, 14 | Errours | Sums of Errours | Sums of Squares of Errours |
| Series (1) | 4, 9, 14 ... | 0, 3, 0 ... | 3 ... | 9 |
| " (2) | 6, 10, 14 ... | 2, 2, 0 ... | 4 ... | 8 |
| " (3) | 5, 10, 15, ... | 1, 2, 1 ... | 4 ... | 6 |

Here, although the first series gives the sum of the errours less than the others, the third series gives the sum of the squares of the errours least; and is therefore, by the proposition on which this Method depends, the *most probable* series of the three.

This Method, in more extensive and complex cases, is a great aid to the calculator in his inferences from facts, and removes much that is arbitrary in the Method of Means.

#### *Sect. IV. The Method of Residues*

15. By either of the preceding Methods we obtain, from observed facts, such Laws as readily offer themselves; and by the Laws thus discovered, the most prominent changes of the observed quantities are accounted for. But in many cases we have, as we have noticed already, *several* Laws of nature operating at the same time, and combining their influences to modify those quantities which are the subjects of observation. In these cases we may, by successive applications of the Methods already pointed out, detect such Laws one after another: but this successive process, though only a repetition of what we have already described, offers some peculiar features which make it convenient to consider it in a separate Section, as the Method of Residues.

16. When we have, in a series of changes of a variable quantity,

discovered *one* Law which the changes follow, detected its Argument, and determined its Magnitude, so as to explain most clearly the course of observed facts, we may still find that the observed changes are not fully accounted for. When we compare the results of our Law with the observations, there may be a difference, or as we may term it, a *Residue,* still unexplained. But this Residue being thus detached from the rest, may be examined and scrutinized in the same manner as the whole observed quantity was treated at first: and we may in this way detect in *it* also a Law of change. If we can do this, we must accommodate this new found Law as nearly as possible to the Residue to which it belongs; and this being done, the difference of our Rule and of the Residue itself, forms a *Second Residue.* This Second Residue we may again bring under our consideration; and may perhaps in *it* also discover some Law of change by which its alterations may be in some measure accounted for. If this can be done, so as to account for a large portion of this Residue, the remaining unexplained part forms a *Third Residue;* and so on.

17. This course has really been followed in various inquiries, especially in those of Astronomy and Tidology. The *Equation of the Center,* for the Moon, was obtained out of the *Residue* of the Longitude, which remained when the *Mean Anomaly* was taken away. This Equation being applied and disposed of, the *Second Residue* thus obtained, gave to Ptolemy the *Evection.* The *Third Residue,* left by the Equation of the Center and the Evection, supplied to Tycho the *Variation* and the *Annual Equation.* And the Residue, remaining from these, has been exhausted by other Equations, of Various arguments, suggested by theory or by observation. In this case, the successive generations of astronomers have gone on, each in its turn executing some step in this Method of Residues. In the examination of the Tides, on the other hand, this method has been applied systematically and at once. The observations readily gave the *Semimensual Inequality;* the *Residue* of this supplied the corrections due to the Moon's *Parallax* and *Declination;* and when these were determined, the *remaining Residue* was explored for the law of the Solar Correction.

18. In a certain degree, the Method of Residues and the Method of Means are *opposite* to each other. For the Method of Residues extricates Laws from their combination, *bringing them into view in suc-*

*cession;* while the Method of Means discovers each Law, not by bringing the others into view, but by *destroying their effect* through an accumulation of observations. By the Method of Residues we should *first* extract the Law of the Parallax Correction of the Tides, and *then,* from the Residue left by this, obtain the Declination Correction. But we might at once employ the Method of Means, and put together all the cases in which the Declination was the same; not allowing for the Parallax in each case, but taking for granted that the Parallaxes belonging to the same Declination would neutralize each other; as many falling above as below the mean Parallax. In cases like this, where the Method of Means is not impeded by a partial coincidence of the Arguments of different unknown Inequalities, it may be employed with almost as much success as the Method of Residues. But still, when the Arguments of the Laws are clearly known, as in this instance, the Method of Residues is more clear and direct, and is the rather to be recommended.

19. If for example, we wish to learn whether the Height of the Barometer exerts any sensible influence on the Height of the Sea's Surface, it would appear that the most satisfactory mode of proceeding, must be to subtract, in the first place, what we know to be the effects of the Moon's Age, Parallax and Declination, and other ascertained causes of change; and to search in the *unexplained Residue* for the effects of barometrical pressure. The contrary course has, however, been adopted, and the effect of the Barometer on the ocean has been investigated by the direct application of the Method of Means, classing the observed heights of the water according to the corresponding heights of the Barometer without any previous reduction. In this manner, the suspicion that the tide of the sea is affected by the pressure of the atmosphere, has been confirmed. This investigation must be looked upon as a remarkable instance of the efficacy of the Method of Means, since the amount of the barometrical effect is much smaller than the other changes from among which it was by this process extricated. But an application of the Method of Residues would still be desirable on a subject of such extent and difficulty.

20. Sir John Herschel, in his *Discourse on the Study of Natural Philosophy* (Articles 158–161), has pointed out the mode of making discoveries by studying Residual Phenomena; and has given several

illustrations of the process. In some of these, he has also considered this method in a wider sense than we have done; treating it as not applicable to quantity only, but to properties and relations of different kinds.

We likewise shall proceed to offer a few remarks on Methods of Induction applicable to other relations than those of quantity.

## CHAPTER VIII
## METHODS OF INDUCTION DEPENDING ON RESEMBLANCE

***Aphorism XLIX***

The Law of Continuity *is this:—that a quantity cannot pass from one amount to another by any change of conditions, without passing through all intermediate magnitudes according to the intermediate conditions. This Law may often be employed to disprove distinctions which have no real foundation.*

***Aphorism L***

The Method of Gradation *consists in taking a number of stages of a property in question, intermediate between two extreme cases which appear to be different. This Method is employed to determine whether the extreme cases are really distinct or not.*

***Aphorism LI***

*The Method of Gradation, applied to decide the question, whether the existing* geological *phenomena arise from existing causes, leads to this result:—That the phenomena do appear to arise from Existing Causes, but that the action of existing causes may, in past times, have transgressed, to any extent, their* recorded *limits of intensity.*

***Aphorism LII***

The Method of Natural Classification *consists in classing cases, not according to any* assumed *Definition, but according to the connexion of the facts themselves, so as to make them the means of asserting general truths.*

*Sect. I. The Law of Continuity*

1. The Law of Continuity is applicable to quantity primarily, and therefore might be associated with the methods treated of in the last chapter: but inasmuch as its inferences are made by a transition from one degree to another among contiguous cases, it will be found to belong more properly to the Methods of Induction of which we have now to speak.

The *Law of Continuity* consists in this proposition,—That a quantity cannot pass from one amount to another by any change of conditions, without passing through all intermediate degrees of magnitude according to the intermediate conditions. And this law may often be employed to correct inaccurate inductions, and to reject distinctions which have no real foundation in nature. For example, the Aristotelians made a distinction between motions according to nature, (as that of a body falling vertically downwards,) and motions contrary to nature, (as that of a body moving along a horizontal plane:) the former, they held, became naturally quicker and quicker, the latter naturally slower and slower. But to this it might be replied, that a horizontal line may pass, by gradual motion, through various inclined positions, to a vertical position: and thus the retarded motion may pass into the accelerated; and hence there must be some inclined plane on which the motion downwards is naturally uniform: which is false, and therefore the distinction of such kinds of motion is unfounded. Again, the proof of the First Law of Motion depends upon the Law of Continuity: for since, by diminishing the resistance to a body moving on a horizontal plane, we diminish the retardation, and this without limit, the law of continuity will bring us at the same time to the case of no resistance and to the case of no retardation.

2. The Law of Continuity is asserted by Galileo in a particular application; and the assertion which it suggests is by him referred to Plato;—namely,[44] that a moveable body cannot pass from rest to a determinate degree of velocity without passing through all smaller degrees of velocity. This law, however, was first asserted in a more general and abstract form by Leibnitz[45]: and was employed by him to show that the laws of motion propounded by Descartes must be false. The Third Cartesian Law of Motion was this[46]: that when one moving body meets another, if the first body have a less momentum than the

second, it will be reflected with its whole motion: but if the first have a greater momentum than the second, it will lose a part of its motion, which it will transfer to the second. Now each of these cases leads, by the Law of Continuity, to the case in which the two bodies have *equal* momentums: but in this case, by the first part of the law the body would *retain all* its motion; and by the second part of the law it would *lose* a portion of it: hence the Cartesian Law is false.

3. I shall take another example of the application of this Law from Professor Playfair's Dissertation on the History of Mathematical and Physical Science.[47] 'The Academy of Sciences at Paris having (in 1724) proposed, as a Prize Question, the Investigation of the Laws of the Communication of Motion, John Bernoulli presented an Essay on the subject very ingenious and profound; in which, however, he denied the existence of hard bodies, because in the collision of such bodies, a finite change of motion must take place in an instant: an event which, on the principle just explained, he maintained to be impossible.' And this reasoning was justifiable: for we can form a *continuous* transition from cases in which the impact manifestly occupies a finite time, (as when we strike a large soft body) to cases in which it is apparently instantaneous. Maclaurin and others are disposed, in order to avoid the conclusion of Bernoulli, to reject the Law of Continuity. This, however, would not only be, as Playfair says, to deprive ourselves of an auxiliary, commonly useful though sometimes deceptive; but what is much worse, to acquiesce in false propositions, from the want of clear and patient thinking. For the Law of Continuity, when rightly interpreted, is *never* violated in actual fact. There are not really any such bodies as have been termed *perfectly hard:* and if we approach towards such cases, we must learn the laws of motion which rule them by attending to the Law of Continuity, not by rejecting it.

4. Newton used the Law of Continuity to suggest, but not to prove, the doctrine of universal gravitation. Let, he said, a terrestrial body be carried as high as the moon: will it not still fall to the earth? and does not the moon fall by the same force?[48] Again: if any one says that there is a material ether which does not gravitate,[49] this kind of matter, by condensation, may be gradually transmuted to the density of the most intensely gravitating bodies: and these gravitating bodies, by taking the internal texture of the condensed ether, may cease to gravi-

tate; and thus the weight of bodies depends, not on their quantity of matter, but on their texture; which doctrine Newton conceived he had disproved by experiment.

5. The evidence of the Law of Continuity resides in the universality of those Ideas, which enter into our apprehension of Laws of Nature. When, of two quantities, one depends upon the other, the Law of Continuity necessarily governs this dependence. Every philosopher has the power of applying this law, in proportion as he has the faculty of apprehending the Ideas which he employs in his induction, with the same clearness and steadiness which belong to the fundamental ideas of Quantity, Space and Number. To those who possess this faculty, the Law is a Rule of very wide and decisive application. Its use, as has appeared in the above examples, is seen rather in the disproof of erroneous views, and in the correction of false propositions, than in the invention of new truths. It is a test of truth, rather than an instrument of discovery.

Methods, however, approaching very near to the Law of Continuity may be employed as positive means of obtaining new truths; and these I shall now describe.

### *Sect. II. The Method of Gradation*

6. To gather together the cases which resemble each other, and to separate those which are essentially distinct, has often been described as the main business of science; and may, in a certain loose and vague manner of speaking, pass for a description of some of the leading procedures in the acquirement of knowledge. The selection of instances which agree, and of instances which differ, in some prominent point or property, are important steps in the formation of science. But when classes of things and properties have been established in virtue of such comparisons, it may still be doubtful whether these classes are separated by distinctions of opposites, or by differences of degree. And to settle such questions, the *Method of Gradation* is employed; which consists in taking intermediate stages of the properties in question, so as to ascertain by experiment whether, in the transition from one class to another, we have to leap over a manifest gap, or to follow a continuous road.

7. Thus for instance, one of the early *Divisions* established by elec-

trical philosophers was that of *Electrics* and *Conductors*. But this division Dr. Faraday has overturned as an essential opposition. He takes[50] a *Gradation* which carries him from Conductors to Non-conductors. Sulphur, or Lac, he says, are held to be non-conductors, but are not rigorously so. Spermaceti is a bad conductor: ice or water better than spermaceti: metals so much better that they are put in a different class. But even in metals the transit of the electricity is not instantaneous: we have in them proof of a retardation of the electric current: "and what reason," Mr. Faraday asks, "why this retardation should not be of the same kind as that in spermaceti, or in lac, or sulphur? But as, in them, retardation is insulation, [and insulation is induction[51]] why should we refuse the same relation to the same exhibitions of force in the metals?"

The process employed by the same sagacious philosopher to show the *identity* of Voltaic and Franklinic electricity, is another example of the same kind.[52] Machine [Franklinic] electricity was made to exhibit the same phenomena as Voltaic electricity, by causing the discharge to pass through a bad conductor, into a very extensive discharging train: and thus it was clearly shown that Franklinic electricity, not so conducted, differs from the other kinds, only in being in a state of successive tension and explosion instead of a state of continued current.

Again; to show that the decomposition of bodies in the Voltaic circuit was not due to the *Attraction* of the Poles,[53] Mr. Faraday devised a beautiful series of experiments, in which these supposed *Poles* were made to assume all possible electrical conditions:—in some cases the decomposition took place against air, which according to common language is not a conductor, nor is decomposed;—in others, against the metallic poles, which are excellent conductors but undecomposable;—and so on: and hence he infers that the decomposition cannot justly be considered as due to the Attraction, or Attractive Powers, of the Poles.

8. The reader of the *Novum Organon* may perhaps, in looking at such examples of the Rule, be reminded of some of Bacon's Classes of Instances, as his *instantiæ absentiæ in proximo*, and his *instantiæ migrantes*. But we may remark that Instances classed and treated as Bacon recommends in those parts of his work, could hardly lead to scientific truth. His processes are vitiated by his proposing to himself the

*form* or *cause* of the property before him, as the object of his inquiry; instead of being content to obtain, in the first place, the *law of phenomena.* Thus his example[54] of a Migrating Instance is thus given. "Let the *Nature inquired into* be that of Whiteness; an Instance Migrating to the production of this property is glass, first whole, and then pulverized; or plain water, and water agitated into a foam; for glass and water are transparent, and not white; but glass powder and foam are white, and not transparent. Hence we must inquire what has happened to the glass or water in that Migration. For it is plain that the *Form of Whiteness* is conveyed and induced by the crushing of the glass and shaking of the water." No real knowledge has resulted from this line of reasoning:—from taking the Natures and Forms of things and of their qualities for the primary subject of our researches.

9. We may easily give examples from other subjects in which the Method of Gradation has been used to establish, or to endeavour to establish, very extensive propositions. Thus Laplace's Nebular Hypothesis,—that systems like our solar system are formed by gradual condensation from diffused masses, such as the nebulæ among the stars,—is founded by him upon an application of this Method of Gradation. We see, he conceives, among these nebulæ instances of all degrees of condensation, from the most loosely diffused fluid, to that separation and solidification of parts by which suns, and satellites, and planets are formed: and thus we have before us instances of systems in all their stages; as in a forest we see trees in every period of growth. How far the examples in this case satisfy the demands of the Method of Gradation, it remains for astronomers and philosophers to examine.

Again; this method was used with great success by Macculloch and others to refute the opinion, put in currency by the Wernerian school of geologists, that the rocks called *trap rocks* must be classed with those to which a *sedimentary* origin is ascribed. For it was shown that a gradual *transition* might be traced from those examples in which trap rocks most resembled stratified rocks, to the lavas which have been recently ejected from volcanoes: and that it was impossible to assign a different origin to one portion, and to the other, of this kind of mineral masses; and as the volcanic rocks were certainly not sedimentary, it followed, that the trap rocks were not of that nature.

Again; we have an attempt of a still larger kind made by Sir C.

Lyell, to apply this Method of Gradation so as to disprove all distinction between the causes by which geological phenomena have been produced, and the causes which are now acting at the earth's surface. He has collected a very remarkable series of changes which have taken place, and are still taking place, by the action of water, volcanoes, earthquakes, and other terrestrial operations; and he conceives he has shown in these a *gradation* which leads, with no wide chasm or violent leap, to the state of things of which geological researches have supplied the evidence.

10. Of the value of this Method on geological speculations, no doubt can be entertained. Yet it must still require a grave and profound consideration, in so vast an application of the Method as that attempted by Sir C. Lyell, to determine what extent we may allow to the steps of our *gradation;* and to decide how far the changes which have taken place in distant parts of the series may exceed those of which we have historical knowledge, without ceasing to be of the *same kind.* Those who, dwelling in a city, see, from time to time, one house built and another pulled down, may say that such *existing causes,* operating through past time, sufficiently explain the existing condition of the city. Yet we arrive at important political and historical truths, by considering the *origin* of a city as an event of a *different order* from those daily changes. The causes which are now working to produce geological results, may be supposed to have been, at some former epoch, so far exaggerated in their operation, that the changes should be paroxysms, not degrees;—that they should violate, not continue, the gradual series. And we have no kind of evidence whether the duration of our historical times is sufficient to give us a just measure of the limits of such degrees;—whether the terms which we have under our notice enable us to ascertain the average rate of progression.

11. The result of such considerations seems to be this:—that we may apply the Method of Gradation in the investigation of geological causes, provided we leave the Limits of the Gradation undefined. But, then, this is equivalent to the admission of the opposite hypothesis: for a continuity of which the successive intervals are not limited, is not distinguishable from discontinuity. The geological sects of recent times have been distinguished as *uniformitarians* and *catastrophists:* the Method of Gradation seems to prove the doctrine of the uniformi-

tarians; but then, at the same time that it does this, it breaks down the distinction between them and the catastrophists.

There are other exemplifications of the use of gradations in Science which well deserve notice: but some of them are of a kind somewhat different, and may be considered under a separate head.

***Sect. III. The Method of Natural Classification***

12. The Method of Natural Classification consists, as we have seen, in grouping together objects, not according to any selected properties, but according to their most important resemblances; and in combining such grouping with the assignation of certain marks of the classes thus formed. The examples of the successful application of this method are to be found in the Classificatory Sciences through their whole extent; as, for example, in framing the Genera of plants and animals. The same method, however, may often be extended to other sciences. Thus the classification of Crystalline Forms, according to their Degree of Symmetry, (which is really an important distinction,) as introduced by Mohs and Weiss, was a great improvement upon Haüy's arbitrary division according to certain assumed primary forms. Sir David Brewster was led to the same distinction of crystals by the study of their optical properties; and the scientific value of the classification was thus strongly exhibited. Mr. Howard's classification of Clouds appears to be founded in their real nature, since it enables him to express the laws of their changes and successions. As we have elsewhere said, the criterion of a true classification is, that it makes general propositions possible. One of the most prominent examples of the beneficial influence of a right classification, is to be seen in the impulse given to geology by the distinction of strata according to the organic fossils which they contain[55]: which, ever since its general adoption, has been a leading principle in the speculations of geologists.

13. The mode in which, in this and in other cases, the Method of Natural Classification directs the researches of the philosopher, is this: —his arrangement being adopted, at least as an instrument of inquiry and trial, he follows the course of the different members of the classification, according to the guidance which Nature herself offers; not prescribing beforehand the marks of each part, but distributing the facts according to the total resemblances, or according to those re-

semblances which he finds to be most important. Thus, in tracing the course of a series of strata from place to place, we identify each stratum, not by any single character, but by all taken together;—texture, colour, fossils, position, and any other circumstances which offer themselves. And if, by this means, we come to ambiguous cases, where different indications appear to point different ways, we decide so as best to preserve undamaged those general relations and truths which constitute the value of our system. Thus although we consider the organic fossils in each stratum as its most important characteristic, we are not prevented, by the disappearance of some fossils, or the addition of others, or by the total absence of fossils, from identifying strata in distant countries, if the position and other circumstances authorize us to do so. And by this Method of Classification, the doctrine of *Geological Equivalents*[56] has been applied to a great part of Europe.

14. We may further observe, that the same method of natural classification which thus enables us to identify strata in remote situations, notwithstanding that there may be great differences in their material and contents, also forbids us to assume the identity of the series of rocks which occur in different countries, when this identity has not been verified by such a continuous exploration of the component members of the series. It would be in the highest degree unphilosophical to apply the special names of the English or German strata to the rocks of India, or America, or even of southern Europe, till it has appeared that in those countries the geological series of northern Europe really exists. In each separate country, the divisions of the formations which compose the crust of the earth must be made out, by applying the Method of Natural Arrangement *to that particular case,* and not by arbitrarily extending to it the nomenclature belonging to another case. It is only by such precautions, that we can ever succeed in obtaining geological propositions, at the same time true and comprehensive; or can obtain any sound general views respecting the physical history of the earth.

15. The Method of Natural Classification, which we thus recommend, falls in with those mental habits which we formerly described as resulting from the study of Natural History. The Method was then termed the *Method of Type,* and was put in opposition to the *Method of Definition.*

The Method of Natural Classification is directly opposed to the process in which we assume and apply *arbitrary* definitions; for in the former Method, we find our classes in nature, and do not make them by marks of our own imposition. Nor can any advantage to the progress of knowledge be procured, by laying down our characters when our arrangements are as yet quite loose and unformed. Nothing was gained by the attempts to *define* Metals by their weight, their hardness, their ductility, their colour; for to all these marks, as fast as they were proposed, exceptions were found, among bodies which still could not be excluded from the list of Metals. It was only when elementary substances were divided into *Natural Classes,* of which classes Metals were one, that a true view of their distinctive characters was obtained. Definitions in the outset of our examination of nature are almost always, not only useless, but prejudicial.

16. When we obtain a Law of Nature by induction from phenomena, it commonly happens, as we have already seen, that we introduce, at the same time, a Proposition and a Definition. In this case, the two are correlative, each giving a real value to the other. In such cases, also, the Definition, as well as the Proposition, may become the basis of rigorous reasoning, and may lead to a series of deductive truths. We have examples of such Definitions and Propositions in the Laws of Motion, and in many other cases.

17. When we have established Natural Classes of objects, we seek for Characters of our classes; and these Characters may, to a certain extent, be called the *Definitions* of our classes. This is to be understood, however, only in a limited sense: for these Definitions are not absolute and permanent. They are liable to be modified and superseded. If we find a case which manifestly belongs to our Natural Class, though violating our Definition, we do not shut out the case, but alter our definition. Thus, when we have made it part of our Definition of the *Rose* family, that they have *alternate stipulate leaves,* we do not, therefore, exclude from the family the genus *Lowæa,* which has *no stipulæ.* In Natural Classifications, our Definitions are to be considered as temporary and provisional only. When Sir C. Lyell established the distinctions of the tertiary strata, which he termed *Eocene, Miocene,* and *Pliocene,* he took a numerical criterion (the proportion of recent species of shells contained in those strata) as the basis of his division.

But now that those kinds of strata have become, by their application to a great variety of cases, a series of Natural Classes, we must, in our researches, keep in view the natural connexion of the formations themselves in different places; and must by no means allow ourselves to be governed by the numerical proportions which were originally contemplated; or even by any amended numerical criterion equally arbitrary; for however amended, Definitions in natural history are never immortal. The etymologies of *Pliocene* and *Miocene* may, hereafter, come to have merely an historical interest; and such a state of things will be no more inconvenient, provided the natural connexions of each class are retained, than it is to call a rock *oolite* or *porphyry*, when it has no roelike structure and no fiery spots.

The Methods of Induction which are treated of in this and the preceding chapter, and which are specially applicable to causes governed by relations of Quantity or of Resemblance, commonly lead us to *Laws of Phenomena* only. Inductions founded upon other ideas, those of Substance and Cause for example, appear to conduct us somewhat further into a knowledge of the essential nature and real connexions of things. But before we speak of these, we shall say a few words respecting the way in which inductive propositions, once obtained, may be verified and carried into effect by their application.

## CHAPTER IX
## OF THE APPLICATION OF INDUCTIVE TRUTHS

***Aphorism LIII***

*When the theory of any subject is established, the observations and experiments which are made in applying the science to use and to instruction, supply a perpetual* verification *of the theory.*

***Aphorism LIV***

*Such observations and experiments, when numerous and accurate, supply also* corrections *of the* constants *involved in the theory; and sometimes, (by the Method of Residues,)* additions *to the theory.*

***Aphorism LV***

*It is worth considering, whether a continued and connected sys-*

*tem of observation and calculation, like that of astronomy, might not be employed with advantage in improving our knowledge of other subjects; as Tides, Currents, Winds, Clouds, Rain, Terrestrial Magnetism, Aurora Borealis, Composition of Crystals, and many other subjects.*

***Aphorism LVI***

*An* extension *of a well-established theory to the explanation of new facts excites admiration as a discovery; but it is a discovery of a lower order than the theory itself.*

***Aphorism LVII***

*The practical inventions which are most important in Art may be either unimportant parts of Science, or results not explained by Science.*

***Aphorism LVIII***

*In modern times, in many departments, Art is constantly guided, governed and advanced by Science.*

***Aphorism LIX***

*Recently several New Arts have been invented, which may be regarded as notable verifications of the anticipations of material benefits to be derived to man from the progress of Science. . . .*

## CHAPTER X
## OF THE INDUCTION OF CAUSES

***Aphorism LX***

*In the* Induction of Causes *the principal Maxim is, that we must be careful to possess, and to apply, with perfect clearness, the Fundamental Idea on which the Induction depends.*

***Aphorism LXI***

*The Induction of Substance, of Force, of Polarity, go beyond mere laws of phenomena, and may be considered as the Induction of Causes.*

***Aphorism LXII***

*The Cause of certain phenomena being inferred, we are led to inquire into the Cause of this Cause, which inquiry must be conducted in the same manner as the previous one; and thus we have the Induction of Ulterior Causes.*

***Aphorism LXIII***

*In contemplating the series of Causes which are themselves the effects of other causes, we are necessarily led to assume a Supreme Cause in the Order of Causation, as we assume a First Cause in Order of Succession. . . .*

# OF THE TRANSFORMATION OF HYPOTHESES IN THE HISTORY OF SCIENCE*

1. The history of science suggests the reflection that it is very difficult for the same person at the same time to do justice to two conflicting theories. Take for example the Cartesian hypothesis of vortices and the Newtonian doctrine of universal gravitation. The adherents of the earlier opinion resisted the evidence of the Newtonian theory with a degree of obstinacy and captiousness which now appears to us quite marvellous: while on the other hand, since the complete triumph of the Newtonians, *they* have been unwilling to allow any merit at all to the doctrine of vortices. It cannot but seem strange, to a calm observer of such changes, that in a matter which depends upon mathematical proofs, the whole body of the mathematical world should pass over, as in this and similar cases they seem to have done, from an opinion confidently held, to its opposite. No doubt this must be, in part, ascribed to the lasting effects of education and early prejudice. The old opinion passes away with the old generation: the new theory grows to its full vigour when its congenital disciples grow to be masters. John Bernoulli continues a Cartesian to the last; Daniel, his son, is a Newtonian from the first. Newton's doctrines are adopted at once in England, for they are the solution of a problem at which his contemporaries have been labouring for years. They find no adherents in France, where Descartes is supposed to have already explained the constitution of the world; and Fontenelle, the secretary of the Academy of Sciences at Paris, dies a Cartesian seventy years after the publication of Newton's *Principia.* This is, no doubt, a part of the explanation of the pertinacity with which opinions are held, both before and after a scientific revolution: but this is not the whole, nor perhaps the most

* From "Of the Transformation of Hypotheses in the History of Science," *Transactions of the Cambridge Philosophical Society,* 9 (May 19, 1851), 139–47.

instructive aspect of the subject. There is another feature in the change, which explains, in some degree, how it is possible that, in subjects, mainly at least mathematical, and therefore claiming demonstrative evidence, mathematicians should hold different and even opposite opinions. And the object of the present paper is to point out this feature in the successions of theories, and to illustrate it by some prominent examples drawn from the history of science.

2. The feature to which I refer is this; that when a prevalent theory is found to be untenable, and consequently, is succeeded by a different, or even by an opposite one, the change is not made suddenly, or completed at once, at least in the minds of the most tenacious adherents of the earlier doctrine; but is effected by a transformation, or series of transformations, of the earlier hypothesis, by means of which it is gradually brought nearer and nearer to the second; and thus, the defenders of the ancient doctrine are able to go on as if still asserting their first opinions, and to continue to press their points of advantage, if they have any, against the new theory. They borrow, or imitate, and in some way accommodate to their original hypothesis, the new explanations which the new theory gives, of the observed facts; and thus they maintain a sort of verbal consistency; till the original hypothesis becomes inextricably confused, or breaks down under the weight of the auxiliary hypotheses thus fastened upon it, in order to make it consistent with the facts.

This often-occuring course of events might be illustrated from the history of the astronomical theory of epicycles and eccentries, as is well known. But my present purpose is to give one or two brief illustrations of a somewhat similar tendency from other parts of scientific history; and in the first place, from that part which has already been referred to, the battle of the Cartesian and Newtonian systems.

3. The part of the Cartesian system of vortices which is most familiarly known to general readers is the explanation of the motions of the planets by supposing them carried round the sun by a kind of whirlpool of fluid matter in which they are immersed: and the explanation of the motions of the satellites round their primaries by similar subordinate whirlpools, turning round the primary, and carried, along with it, by the primary vortex. But it should be borne in mind that a part of the Cartesian hypothesis which was considered quite as

important as the cosmical explanation, was the explanation which it was held to afford of terrestrial gravity. Terrestrial gravity was asserted to arise from the motion of the vortex of subtle matter which revolved round the earth's axis and filled the surrounding space. It was maintained that by the rotation of such a vortex, the particles of the subtle matter would exert a centrifugal force, and by virtue of that force, tend to recede from the center: and it was held that all bodies which were near the earth, and therefore immersed in the vortex, would be pressed towards the center by the effort of the subtle matter to recede from the center.[1]

These two assumed effects of the Cartesian vortices—to carry bodies in their stream, as straws are carried round by a whirlpool, and to press bodies to the center by the centrifugal effort of the whirling matter—must be considered separately, because they were modified separately, as the progress of discussion drove the Cartesians from point to point. The former effect indeed, the *dragging* force of the vortex, as we may call it, would not bear working out on mechanical principles at all; for as soon as the law of motion was acknowledged (which Descartes himself was one of the loudest in proclaiming), that a body in motion keeps all the motion which it has, and receives in addition all that is impressed upon it; as soon, in short, as philosophers rejected the notion of an inertness in matter which constantly retards its movements,—it was plain that a planet perpetually dragged onwards in its orbit by a fluid moving quicker than itself, must be perpetually accelerated; and therefore could not follow those constantly-recurring cycles of quicker and slower motion which the planets exhibit to us.

The Cartesian mathematicians, then, left untouched the calculation of the progressive motion of the planets; and, clinging to the assumption that a vortex would produce a tendency of bodies to the center, made various successive efforts to construct their vortices in such a manner that the centripetal forces produced by them should coincide with those which the phenomena required, and therefore of course, in the end, with those which the Newtonian theory asserted.

In truth, the Cartesian vortex was a bad piece of machinery for producing a central force: from the first, objections were made to the sufficiency of its mechanism, and most of these objections were very unsatisfactorily answered, even granting the additional machinery which its defenders demanded. One formidable objection was soon started, and continued to the last to be the torment of the Cartesians. If terrestrial gravity, it was urged, arise from the centrifugal force of a vortex which revolves about the earth's axis, terrestrial gravity ought to act in planes perpendicular to the earth's axis, instead of tending to the earth's center. This objection was taken by James Bernoulli,[2] and by Huyghens[3] not long after the publication of Descartes's *Principia.* Huyghens (who adopted the theory of vortices with modifications of his own) supposes that there are particles of the fluid matter which move about the earth in every possible direction, within the spherical space which includes terrestrial objects; and that the greater part of these motions being in spherical surfaces concentric with the earth, produces a tendency towards the earth's center.

This was a procedure tolerably arbitrary, but it was the best which could be done. Saurin, a little later,[4] gave nearly the same solution of this difficulty. The solution, identifying a vortex of some kind with a central force, made the hypothesis of vortices applicable wherever central forces existed; but then, in return, it deprived the image of a vortex of all that clearness and simplicity which had been its first great recommendation.

But still there remained difficulties not less formidable. According to this explanation of gravity, since the tendency of bodies to the earth's center arose from the superior centrifugal force of the whirling matter which pushed them inward as water pushes a light body upward, bodies ought to tend more strongly to the center in proportion as they are less dense. The rarest bodies should be the heaviest; contrary to what we find.

Descartes's original solution of this difficulty has a certain degree of ingenuity. According to him (*Princip.* IV. 23) a terrestrial body consists of particles of the *third element,* and the more it has of such particles, the more it excludes the parts of the *celestial matter,* from the revolution of which matter gravity arises; and therefore the denser is the terrestrial body, and the heavier it will be.

But though this might satisfy him, it could not satisfy the mathematicians who followed him, and tried to reduce his system to calculation on mechanical principles. For how could they do this, if the celestial matter, by the operation of which the phenomena of force and motion were produced, was so entirely different from ordinary matter, which alone had supplied men with experimental illustrations of mechanical principles? In order that the celestial matter, by its whirling, might produce the gravity of heavy bodies, it was mechanically necessary that it must be very dense; and *dense* in the ordinary sense of the term; for it was by regarding density in the ordinary sense of the term that the mechanical necessity had been established.

The Cartesians tried to escape this result (Huyghens, *Pesanteur*, p. 161, and John Bernoulli, *Nouvelles Pensées*, Art. 31) by saying that there were two meanings of *density* and *rarity;* that some fluids might be rare by having their particles far asunder, others, by having their particles very small though in contact. But it is difficult to think that they could, as persons well acquainted with mechanical principles, satisfy themselves with this distinction; for they could hardly fail to see that the mechanical effect of any portion of fluid depends upon the total mass moved, not on the size of its particles.

Attempts made to exemplify the vortices experimentally only showed more clearly the force of this difficulty. Huyghens had found that certain bodies immersed in a whirling fluid tended to the center of the vortex. But when Saulmon[5] a little later made similar experiments, he had the mortification of finding that the heaviest bodies had the greatest tendency to recede from the axis of the vortex. "The result is," as the Secretary of the Academy (Fontenelle) says, "exactly the opposite of what we could have wished, for the [Cartesian] system of gravity: but we are not to despair; sometimes in such researches disappointment leads to ultimate success."

But, passing by this difficulty, and assuming that in some way or other a centripetal force arises from the centrifugal force of the vortex, the Cartesian mathematicians were naturally led to calculate the circumstances of the vortex on mechanical principles; especially Huyghens, who had successfully studied the subject of centrifugal force. Accordingly, in his little treatise on the *Cause of Gravitation* (p. 143), he calculates the velocity of the fluid matter of the vortex, and finds that, at a point in the equator, it is 17 times the velocity of the earth's rotation.

It may naturally be asked, how it comes to pass that a stream of fluid, dense enough to produce the gravity of bodies by its centrifugal force, moving with a velocity 17 times that of the earth (and therefore moving round the earth in 85 minutes), does not sweep all terrestrial objects before it. But to this Huyghens had already replied (p. 137), that there are particles of the fluid moving *in all directions*, and therefore that they neutralize each other's action, so far as lateral motion is concerned.

And thus, as early as this treatise of Huyghens, that is, in three years from the publication of Newton's *Principia*, a vortex is made to mean nothing more than some machinery or other for producing a central force. And this is so much the case, that Huyghens commends (p. 165), as confirming his own calculation of the velocity of his vortex, Newton's proof that at the Moon's orbit the centripetal force is equal to the centrifugal; and that thus, this force is less than the centripetal force at the earth's surface in the inverse proportion of the squares of the distances.

John Bernoulli, in the same manner, but with far less clearness and less candour, has treated the hypothesis of vortices as being principally a hypothetical cause of central force. He had repeated occasions given him of propounding his inventions for propping up the Cartesian doctrine, by the subjects proposed for prizes by the Paris Academy of Sciences; in which competition Cartesian speculations were favorably received. Thus the subject of the Prize Essays for 1730 was, the explanation of the Elliptical Form of the planetary orbits and of the Motion of their Aphelia, and the prize was assigned to John Bernoulli, who gave the explanation on Cartesian principles. He explains the elliptical figure, not as Descartes himself had done, by supposing the vortex which carries the planet round the sun to be itself squeezed into an elliptical form by the pressure of contiguous vortices; but he supposes the planet, while it is carried round by the vortex, to have a limited oscillatory motion to and from the center, produced by its being originally, not at the distance at which it would float in equilibrium in the vortex, but above or below that point. On this supposition, the planet would oscillate to and from the center, Bernoulli says, like the mercury when deranged in a barometer: and it is evident that such an oscillation, combined with a motion round the center, might produce an oval curve, either with a fixed or with a moveable aphelion.

All this however merely amounts to a possibility that the oval *may* be an ellipse, not to a proof that it will be so; nor does Bernoulli advance further.

It was necessary that the vortices should be adjusted in such a manner as to account for Kepler's laws; and this was to be done by making the velocity of each stratum of the vortex depend in a suitable manner on its radius. The Abbé de Molières attempted this on the supposition of elliptical vortices, but could not reconcile Kepler's first two laws, of equal elliptical areas in equal times, with his third law, that the squares of the periodic times are as the cubes of the mean distances.[6] Bernoulli, with his circular vortices, could accommodate the velocities at different distances so that they should explain Kepler's laws. He pretended to prove that Newton's investigations respecting vortices (in the ninth Section of the Second Book of the *Principia*) were mechanically erroneous; and in truth, it must be allowed that, besides several arbitrary assumptions, there are some errors of reasoning in them. But for the most part, the more enlightened Cartesians were content to accept Newton's account of the motions and forces of the solar system as part of their scheme; and to say only that the hypothesis of vortices explained the origin of the Newtonian forces; and that thus theirs was a philosophy of a higher kind. Thus it is asserted (Mém. Acad. 1734), that M. de Molières retains the beautiful theory of Newton entire, only he renders it in a sort less Newtonian, by disentangling it from attraction, and transferring it from a vacuum into a plenum. This plenum, though not its native region, frees it from the need of attraction, which is all the better for it. These points were the main charms of the Cartesian doctrine in the eyes of its followers; —the getting rid of attractions, which were represented as a revival of the Aristotelian "occult qualities," "substantial forms," or whatever else was the most disparaging way of describing the bad philosophy of the dark ages[7];—and the providing some material intermedium, by means of which a body may affect another at a distance; and thus avoid the reproach urged against the Newtonians, that they made a body act where it was not. And we are the less called upon to deny that this last feature in the Newtonian theory was a difficulty, inasmuch as Newton himself was never unwilling to allow that gravity might be merely an effect produced by some ulterior cause.

With such admissions on the two sides, it is plain that the Newtonian and Cartesian systems would coincide, if the hypothesis of vortices could be modified in such a way as to produce the force of gravitation. All attempts to do this, however, failed: and even John Bernoulli, the most obstinate of the mathematical champions of the vortices, was obliged to give them up. In his Prize Essay for 1734, (on the Inclinations of the Planetary Orbits[8]), he says (Art. VIII.), "The gravitation of the Planets towards the center of the Sun and the weight of bodies towards the center of the earth has not, for its cause, either the attraction of M. Newton, or the centrifugal force of the matter of the vortex according to M. Descartes;" and he then goes on to assert that these forces are produced by a perpetual torrent of matter tending to the center on all sides, and carrying all bodies with it. Such a hypothesis is very difficult to refute. It has been taken up in more modern times by Le Sage,[9] with some modifications; and may be made to account for the principal facts of the universal gravitation of matter. The great difficulty in the way of such a hypothesis is, the overwhelming thought of the whole universe filled with torrents of an invisible but material and tangible substance, rushing in every direction in infinitely prolonged straight lines and with immense velocity. Whence can such matter come, and whither can it go? Where can be its perpetual and infinitely distant fountain, and where the ocean into which it pours itself when its infinite course is ended? A revolving whirlpool is easily conceived and easily supplied; but the central torrent of Bernoulli, the infinite streams of particles of Le Sage, are an explanation far more inconceivable than the thing explained.

But however the hypothesis of vortices, or some hypothesis substituted for it, was adjusted to explain the facts of attraction to a center, this was really nearly all that was meant by a vortex or a "tourbillon," when the system was applied. Thus in the case of the last act of homage to the Cartesian theory which the French Academy rendered in the distribution of its prizes, the designation of a Cartesian Essay in 1741 (along with three Newtonian ones) as worthy of a prize for an explanation of the Tides; the difference of high and low water was not explained, as Descartes has explained it, by the pressure, on the ocean, of the terrestrial vortex, forced into a strait where it passes under the Moon; but the waters were supposed to rise towards the Moon, the terrestrial vortex being disturbed and broken by the Moon, and therefore less effective in forcing them down. And in giving an account of a Tourmaline from Ceylon (Acad. Sc. 1717), when it has been ascertained that it attracts and repels substances, the writer adds, as a matter of course, "It would seem that it has a vortex." As another example, the elasticity of a body was ascribed to vortices between its particles: and in general, as I have said, a vortex implied what we now imply by speaking of a central force.

4. In the same manner vortices were ascribed to the Magnet, in order to account for its attractions and repulsions. But we may note a circumstance which gave a special turn to the hypothesis of vortices as applied to this subject, and which may serve as a further illustration of the manner in which a transition may be made from one to the other of two rival hypotheses.

If iron filings be brought near a magnet, in such a manner as to be at liberty to assume the position which its polar action assigns to them; (for instance, by strewing them upon a sheet of paper while the two poles of the magnet are close below the paper;) they will arrange themselves in certain curves, each proceeding from the N. to the S. pole of the magnet, like the meridians in a map of the globe. It is easily shown, on the supposition of magnetic attraction and repulsion, that these *magnetic curves,* as they are termed, are each a curve whose tangent at every point is the direction of a small line or particle, as determined by the attraction and repulsion of the two poles. But if we suppose a *magnetic vortex* constantly to flow out of one pole and into the other, in streams which follow such curves, it is evident that such a vortex, being supposed to exercise material pressure and impulse, would arrange the iron filings in corresponding streams, and would thus produce the phenomenon which I have described. And the hypothesis of *central torrents* of Bernoulli or Le Sage which I have referred to, would, in its application to magnets, really become this hypothesis of a magnetic vortex, if we further suppose that the matter of the torrents which proceed to one pole and from the other, mingles its streams, so as at each point to produce a stream in the resulting direction. Of course we shall have to suppose two sets of magnetic torrents; —a boreal torrent, proceeding to the north pole, and from the south pole of a magnet; and an austral torrent proceeding to the south and from the north pole:—and with these suppositions, we make a transition from the hypothesis of attraction and repulsion, to the Cartesian hypothesis of vortices, or at least, torrents, which determine bodies to their magnetic positions by impulse.

Of course it is to be expected that, in this as in the other case, when we follow the hypothesis of impulse into detail, it will need to be loaded with so many subsidiary hypotheses, in order to accommodate it to the phenomena, that it will no longer seem tenable. But the plausibility of the hypothesis in its first application cannot be denied:—for, it may be observed, the two *opposite* streams would counteract each other so as to produce no local *motion*, only *direction*. And this case may put us on our guard against other suggestions of forces acting in curve lines, which may at first sight appear to be discerned in magnetic and electric phenomena. Probably such curve lines will all be found to be only resulting lines, arising from the direct action and combination of elementary attraction and repulsion.

5. There is another case in which it would not be difficult to devise a mode of transition from one to the other of two rival theories; namely, in the case of the emission theory and the undulation theory of Light. Indeed several steps of such a transition have already appeared in the history of optical speculation; and the conclusive objection to the emission theory of light, as to the Cartesian theory of vortices, is, that no amount of additional hypotheses will reconcile it to the phenomena. Its defenders had to go on adding one piece of machinery after another, as new classes of facts came into view, till it became more complex and unmechanical than the theory of epicycles and eccentrics at its worst period. Otherwise, as I have said, there was nothing to prevent the emission theory from migrating into the undulatory theory, and as the theory of vortices did into the theory of attraction. For the emissionists allow that rays may *interfere;* and that these interferences may be modified by alternate *fits* in the rays; now these fits are already a kind of *undulation*. Then again the phenomena of polarized light show that the fits or undulations must have a *transverse* character: and there is no reason why emitted rays should not be subject to *fits* of *transverse* modification as well as to any other fits. In short, we may add to the emitted rays of the one theory, all the properties which belong to the undulations of the other, and thus ac-

count for all the phenomena on the emission theory; with this limitation only, that the emission will have no share in the explanation, and the undulations will have the whole. If, instead of conceiving the universe full of a *stationary* ether, we suppose it to be full of etherial particles moving in every direction; and if we suppose in the one case and in the other, this ether to be susceptible of undulations proceeding from every luminous point; the results of the two hypotheses will be the same; and all we shall have to say is, that the supposition of the emissive motion of the particles is superfluous and useless.

6. This view of the manner in which rival theories pass into one another appears to be so unfamiliar to those who have only slightly attended to the history of science, that I have thought it might be worth while to illustrate it by a few examples.

It might be said, for instance, by such persons,[10] "Either the planets are not moved by vortices, or they do not move by the law by which heavy bodies fall. It is impossible that both opinions can be true." But it appears, by what has been said above, that the Cartesians did hold both opinions to be true; and one with just as much reason as the other, on their assumptions. It might be said in the same manner, "Either it is false that the planets are made to describe their orbits by the above quasi-Cartesian theory of Bernoulli, or it is false that they obey the Newtonian theory of gravitation." But this would be said quite erroneously; for if the hypothesis of Bernoulli be true, it is so because it agrees in its result with the theory of Newton. It is not only possible that both opinions may be true, but it is certain that if the first be so, the second is. It might be said again, "Either the planets describe their orbits by an inherent virtue, or according to the Newton theory." But this again would be erroneous, for the Newtonian doctrine decided nothing as to whether the force of gravitation was inherent or not. Cotes held that it was, though Newton strongly protested against being supposed to hold such an opinion. The word *inherent* is no part of the physical theory, and will be asserted or denied according to our metaphysical views of the essential attributes of matter and force.

Of course, the possibility of two rival hypotheses being true, one of which takes the explanation a step higher than the other, is not affected by the impossibility of two contradictory assertions of the *same order* of generality being both true. If there be a new-discovered comet, and if one astronomer asserts that it will return once in *every* twenty years, and another, that it will return once in every thirty years, both cannot be right. But if an astronomer says that though its interval was in the last instance 30 years, it will only be 20 years to the next return, in consequence of perturbation and resistance, he may be perfectly right.

And thus, when different and rival explanations of the same phenomena are held, till one of them, though long defended by ingenious men, is at last driven out of the field by the pressure of facts, the defeated hypothesis is transformed before it is extinguished. Before it has disappeared, it has been modified so as to have all palpable falsities squeezed out of it, and subsidiary provisions added, in order to reconcile it with the phenomena. It has, in short, been penetrated, infiltrated, and metamorphosed by the surrounding medium of truth, before the merely arbitrary and erroneous residuum has been finally ejected out of the body of permanent and certain knowledge.

Chapter *iv*

# Whewell's Reply to Mill

# MR. MILL'S LOGIC*

The *History of the Inductive Sciences* was published in 1837, and the *Philosophy of the Inductive Sciences* in 1840. In 1843 Mr. Mill published his *System of Logic*,[1] in which he states that without the aid derived from the facts and ideas in my volumes, the corresponding portion of his own would most probably not have been written, and quotes parts of what I have said with commendation. He also, however, dissents from me on several important and fundamental points, and argues against what I have said thereon. I conceive that it may tend to bring into a clearer light the doctrines which I have tried to establish, and the truth of them, if I discuss some of the differences between us, which I shall proceed to do.

Mr. Mill's work has had, for a work of its abstruse character, a circulation so extensive, and admirers so numerous and so fervent, that it needs no commendation of mine. But if my main concern at present had not been with the points in which Mr. Mill *differs* from me, I should have had great pleasure in pointing out passages, of which there are many, in which Mr. Mill appears to me to have been very happy in promoting or in expressing philosophical truth.

There is one portion of his work indeed which tends to give it an interest of a wider kind than belongs to that merely scientific truth to which I purposely and resolutely confined my speculations in the works to which I have referred. Mr. Mill has introduced into his work a direct and extensive consideration of the modes of dealing with moral and political as well as physical questions; and I have no doubt that this part of his book has, for many of his readers, a more lively interest than any other. Such a comprehensive scheme seems to give to doctrines respecting science a value and a purpose which they cannot

* From *Of Induction, with Especial reference to Mr. J. Stuart Mill's System of Logic* (London 1849).

have, so long as they are restricted to mere material sciences. I still retain the opinion, however, upon which I formerly acted, that the philosophy of science is to be extracted from the portions of science, which are universally allowed to be most certainly established, and that those are the physical sciences. I am very far from saying, or thinking, that there is no such thing as Moral and Political Science, or that no method can be suggested for its promotion; but I think that by attempting at present to include the Moral sciences in the same formulae with the Physical, we open far more controversies than we close; and that in the moral as in the physical sciences, the first step towards showing how truth is to be discovered, is to study some portion of it which is assented to so as to be beyond controversy.

### *I. What is Induction?*

1. Confining myself, then, to the material sciences, I shall proceed to offer my remarks on Induction with especial reference to Mr. Mill's work. And in order that we may, as I have said, proceed as intelligibly as possible, let us begin by considering what we mean by *Induction*, as a mode of obtaining truth; and let us note whether there is any difference between Mr. Mill and me on this subject.

"For the purposes of the present inquiry," Mr. Mill says (i. 347), "Induction may be defined the operation of discovering and forming general propositions:" meaning, as appears by the context, the discovery of them from particular facts. He elsewhere (i. 370) terms it "generalization from experience:" and again he speaks of it with greater precision as the inference of a more general proposition from less general ones.

2. Now to these definitions and descriptions I assent as far as they go; though, as I shall have to remark, they appear to me to leave unnoticed, a feature which is very important, and which occurs in all cases of Induction, so far as we are concerned with it. Science, then, consists of general propositions, inferred from particular facts, or from less general propositions, by Induction; and it is our object to discern the nature and laws of *Induction* in this sense. That the propositions are general, or are more general than the facts from which they are inferred, is an indispensable part of the notion of Induction, and is essential to any discussion of the process, as the mode of arriving at Science, that is, at a body of general truths.

3. I am obliged therefore to dissent from Mr. Mill when he includes, in his notion of Induction, the process by which we arrive *at individual facts* from other facts *of the same order of particularity.*

Such inference is, at any rate, not Induction *alone;* if it be Induction at all, it is Induction applied to an example.

For instance, it is a general law, obtained by Induction from particular facts, that a body falling vertically downwards from rest, describes spaces proportional to the squares of the times. But that a particular body will fall through 16 feet in one second and 64 feet in two seconds, is not an induction simply, it is a result obtained by applying the inductive law to a particular case.

But further, such a process is often not induction *at all.* That a ball striking another ball directly will communicate to it as much momentum as the striking ball itself loses, is a law established by induction: but if, from habit or practical skill, I make one billiard-ball strike another, so as to produce the velocity which I wish, without knowing or thinking of the general law, the term *Induction* cannot then be rightly applied. If I *know the law* and act upon it, I have in my mind both the general induction and its particular application. But if I act by the ordinary billiard-player's skill, without thinking of momentum or law, there is no Induction in the case.

4. This distinction becomes of importance, in reference to Mr. Mill's doctrine, because he has extended his use of the term *Induction,* not only to the cases in which the general induction is consciously applied to a particular instance; but to the cases in which the particular instance is dealt with by means of experience, in that rude sense in which *experience* can be asserted of brutes; and in which, of course, we can in no way imagine that the law is possessed or understood, as a general proposition. He has thus, as I conceive, overlooked the broad and essential difference between speculative knowledge and practical action; and has introduced cases which are quite foreign to the idea of science, alongside with cases from which we may hope to obtain some views of the nature of science and the processes by which it must be formed.

5. Thus (ii. 232) he says, "This inference of one particular fact from another is a case of induction. It is of this sort of induction that brutes are capable." And to the same purpose he had previously said (i. 251), "He [the burnt child who shuns the fire] is not generalizing:

he is inferring a particular from particulars. In the same way also, brutes reason . . . not only the burnt child, but the burnt dog, dreads the fire."

6. This confusion, (for such it seems to me,) of knowledge with practical tendencies, is expressed more in detail in other places. Thus he says (i. 118), "I cannot dig the ground unless I have an idea of the ground and of a spade, and of all the other things I am operating upon."

7. This appears to me to be a use of words which can only tend to confuse our idea of knowledge by obliterating all that is distinctive in *human* knowledge. It seems to me quite false to say that I cannot dig the ground, unless I have an idea of the ground and of my spade. Are we to say that we cannot *walk* the ground, unless we have an idea of the ground, and of our feet, and of our shoes, and of the muscles of our legs? Are we to say that a mole cannot dig the ground, unless he has an idea of the ground and of the snout and paws with which he digs it? Are we to say that a pholas cannot perforate a rock, unless he have an idea of the rock, and of the acid with which he corrodes it?

8. This appears to me, as I have said, to be a line of speculation which can lead to nothing but confusion. The knowledge concerning which I wish to inquire is *human* knowledge. And in order that I may have any chance of success in the inquiry, I find it necessary to single out that kind of knowledge which is especially and distinctively human. Hence, I pass by, in this part of my investigation, all the *knowledge,* if it is to be so called, which man has in no other way than brutes have it;—all that merely shows itself in action. For though action may be modified by habit, and habit by experience, in animals as well as in men, such experience, so long as it retains that merely practical form, is no part of the materials of science. Knowledge in a *general* form, is alone knowledge for that purpose; and to *that,* therefore, I must confine my attention; at least till I have made some progress in ascertaining its nature and laws, and am thus prepared to compare such knowledge,—*human knowledge* properly so called,—with mere animal tendencies to action; or even with practical skill which does not include, as for the most part practical skill does not include, speculative knowledge.

9. And thus, I accept Mr. Mill's definition of Induction only in its

first and largest form; and reject, as useless and mischievous for our purposes, his extension of the term to the practical influence which experience of one fact exercises upon a creature dealing with similar facts. Such influence cannot be resolved into *ideas* and *induction,* without, as I conceive, making all our subsequent investigation vague and heterogeneous, indefinite and inconclusive. If we must speak of animals as *learning* from experience, we may at least abstain from applying to them terms which imply that they learn, in the same way in which men learn astronomy from the stars, and chemistry from the effects of mixture and heat. And the same may be said of the language which is to be used concerning what *men* learn, when their *learning* merely shows itself in action, and does not exist as a general thought. *Induction* must not be applied to such cases. *Induction* must be confined to cases where we have in our minds general propositions, in order that the sciences, which are our most instructive examples of the process we have to consider, may be, in any definite and proper sense, *Inductive* Sciences.

10. Perhaps some persons may be inclined to say that this difference of opinion, as to the extent of meaning which is to be given to the term *Induction,* is a question merely of words; a matter of definition only. This is a mode in which men in our time often seem inclined to dispose of philosophical questions; thus evading the task of forming an opinion upon such questions, while they retain the air of looking at the subject from a more comprehensive point of view. But as I have elsewhere said, such questions of definition are never questions of definition merely. A proposition is always implied along with the definition; and the truth of the proposition depends upon the settlement of the definition. This is the case in the present instance. We are speaking of *Induction,* and we mean that kind of Induction by which the sciences now existing among men have been constructed. On this account it is, that we cannot include, in the meaning of the term, mere practical tendencies or practical habits; for science is not constructed of these. No accumulation of these would make up any of the acknowledged sciences. The elements of such sciences are something of a kind different from practical habits. The elements of such sciences are principles which we *know;* truths which can be contemplated as being *true.* Practical habits, practical skill, instincts and the like, appear in action,

and in action only. Such endowments or acquirements show themselves when the occasion for action arrives, and then, show themselves in the act; without being put, or being capable of being put, in the form of truths contemplated by the intellect. But the elements and materials of Science are necessary truths contemplated by the intellect. It is by consisting of such elements and such materials, that Science *is* Science. Hence a use of the term *Induction* which requires us to obliterate this distinction, must make it impossible for us to arrive at any consistent and intelligible view of the nature of Science, and of the mental process by which Sciences come into being. We must, for the purpose which Mr. Mill and I have in common, retain his larger and more philosophical definition of Induction,—that it is the inference of a more general proposition from less general ones.

11. Perhaps, again, some persons may say, that practical skill and practical experience *lead to* science, and may therefore be included in the term *Induction,* which describes the formation of science. But to this we reply, that these things lead to science as occasions only, and do not form part of science; and that science begins then only when we look at the facts in a general point of view. This distinction is essential to the philosophy of science. The rope-dancer may, by his performances, suggest, to himself or to others, properties of the center of gravity; but this is so, because man has a tendency to speculate and to think of general truths, as well as a tendency to dance on a rope on special occasions, and to acquire skill in such dancing by practice. The rope-dancer does not dance by Induction, any more than the dancing dog does. To apply the terms Science and Induction to such cases, carries us into the regions of metaphor; as when we call birds of passage "wise meteorologists," or the bee "a natural chemist, who turns the flower-dust into honey." This is very well in poetry: but for our purposes we must avoid recognizing these cases as really belonging to the sciences of meteorology and chemistry,—as really cases of Induction. Induction for us is general propositions, *contemplated as such,* derived from particulars.

Science may result *from* experience and observation *by* Induction; but Induction is not therefore the same thing as experience and observation. Induction is experience or observation *consciously* looked at in a *general* form. This consciousness and generality are necessary parts of that knowledge which is science. And accordingly, on the other hand, science cannot result from mere Instinct, as distinguished from Reason; because Instinct by its nature is not conscious and general, but operates blindly and unconsciously in particular cases, the actor not seeing or thinking of the rule which he obeys.

12. A little further on I shall endeavour to show that not only a general *thought,* but a general *word* or phrase is a requisite element in Induction. This doctrine, of course, still more decidedly excludes the case of animals, and of mere practical knowledge in man. A burnt child dreads the fire; but reason must be unfolded, before the child learns to understand the words "fire will hurt you." The burnt dog never thus learns to understand words. And this difference points to an entirely different state of thought in the two cases: or rather, to a difference between a state of rational thought on the one hand, and of mere practical instinct on the other.

13. Besides this difference of speculative thought and practical instinct which thus are, as appears to me, confounded in Mr. Mill's philosophy, in such a way as tends to destroy all coherent views of human knowledge, there is another set of cases to which Mr. Mill applies the term *Induction,* and to which it appears to me to be altogether inapplicable. He employs it to describe the mode in which superstitious men, in ignorant ages, were led to the opinion that striking natural events presaged or accompanied calamities. Thus he says (i. 389), "The opinion so long prevalent that a comet or any other unusual appearance in the heavenly regions was the precursor of calamities to mankind, or at least to those who witnessed it; the belief in the oracles of Delphi and Dodona; the reliance on astrology, or on the weather-prophecies in almanacs; were doubtless inductions supposed to be grounded on experience;" and he speaks of these insufficient inductions being extinguished by the stronger inductions subsequently obtained by scientific inquiry. And in like manner, he says in another place (i. 367), "Let us now compare different predictions: the first, that eclipses will occur whenever one planet or satellite is so situated as to cast its shadow upon another: the second, that they will occur whenever some great calamity is impending over mankind."

14. Now I cannot see how anything but confusion can arise from applying the term *Induction* to superstitious fancies like those here

mentioned. They are not imperfect truths, but entire falsehoods. Of that, Mr. Mill and I are agreed: how then can they exemplify the progress towards truth? They were not collected from the facts by seeking a law of their occurrence; but were suggested by an imagination of the anger of superior powers shown by such deviations from the ordinary course of nature. If we are to speak of *inductions* to any purpose, they must be such inductions as represent the facts, in some degree at least. It is not meant, I presume, that these opinions are in any degree true: to what purpose then are they adduced? If I were to hold that my dreams predict or conform to the motions of the stars or of the clouds, would this be an induction? It would be so, as much one as those here so denominated: yet what but confusion could arise from classing it among scientific truths? Mr. Mill himself has explained (ii. 389) the way in which such delusions as the prophecies of almanac-makers, and the like, obtain credence; namely, by the greater effect which the positive instances produce on ordinary minds in comparison with the negative, when the rule has once taken possession of their thoughts. And this being, as he says, the recognized explanation of such cases, why should we not leave them to their due place, and not confound and perplex the whole of our investigation by elevating them to the rank of "inductions"? The very condemnation of such opinions is that they are not at all inductive. When we have made any progress in our investigation of the nature of science, to attempt to drive us back to the wearisome discussion of such elementary points as these, is to make progress hopeless.

### *II. Induction or Description?*

15. In the cases hitherto noticed, Mr. Mill extends the term *Induction,* as I think, too widely, and applies it to cases to which it is not rightly applicable. I have now to notice a case of an opposite kind, in which he does not apply it where I do, and condemns me for using it in such a case. I had spoken of Kepler's discovery of the Law, that the planets move round the sun in ellipses, as an example of Induction. The separate facts of any planet (Mars, for instance,) being in certain places at certain times, are all included in the general proposition which Kepler discovered, that Mars describes an ellipse of a certain form and position. This appears to me a very simple but a very dis-

tinct example of the operation of discovering general propositions; general, that is, with reference to particular facts; which operation Mr. Mill, as well as myself, says is Induction. But Mr. Mill denies this operation in this case to be Induction at all (i. 357). I should not have been prepared for this denial by the previous parts of Mr. Mill's book, for he had said just before (i. 350), "such facts as the magnitudes of the bodies of the solar system, their distances from each other, the figure of the earth and its rotation . . . are proved indirectly, by the aid of inductions founded on other facts which we can more easily reach." If the figure of the earth and its rotation are proved by Induction, it seems very strange, and is to me quite incomprehensible, how the figure of the earth's orbit and its revolution (and of course, of the figure of Mars's orbit and his revolution in like manner,) are not also proved by Induction. No, says Mr. Mill, Kepler, in putting together a number of places of the planet into one figure, only performed an act of *description*. "This descriptive operation," he adds (i. 359), "Mr. Whewell, by an aptly chosen expression, has termed Colligation of Facts." He goes on to commend my observations concerning this process, but says that, according to the old and received meaning of the term, it is not Induction at all.

16. Now I have already shown that Mr. Mill himself, a few pages earlier, had applied the term *Induction* to cases undistinguishable from this in any essential circumstance. And even in this case, he allows that Kepler did really perform an act of Induction (i. 358), "namely, in concluding that, because the observed places of Mars were correctly represented by points in an imaginary ellipse, therefore Mars would continue to revolve in that same ellipse; and even in concluding that the position of the planet during the time which had intervened between the two observations must have coincided with the intermediate points of the curve." Of course, in Kepler's Induction, of which I speak, I include all this; all this is included in speaking of the *orbit* of Mars: a continuous line, a periodical motion, are implied in the term *orbit*. I am unable to see what would remain of Kepler's discovery, if we take from it these conditions. It would not only not be an induction, but it would not be a description, for it would not recognize that Mars moved in an orbit. Are particular positions to be conceived as points in a curve, without thinking of the intermediate po-

sitions as belonging to the same curve? If so, there is no law at all, and the facts are not bound together by any intelligible tie.

In another place (ii. 209) Mr. Mill returns to his distinction of Description and Induction; but without throwing any additional light upon it, so far as I can see.

17. The only meaning which I can discover in this attempted distinction of Description and Induction is, that when particular facts are bound together by their relation in *space*, Mr. Mill calls the discovery of the connexion *Description*, but when they are connected by other general relations, as time, cause and the like, Mr. Mill terms the discovery of the connexion *Induction*. And this way of making a distinction, would fall in with the doctrine of other parts of Mr. Mill's book, in which he ascribes very peculiar attributes to space and its relations, in comparison with other Ideas, (as I should call them). But I cannot see any ground for this distinction, of connexion according to space and other connexions of facts.

To stand upon such a distinction, appears to me to be the way to miss the general laws of the formation of science. For example: The ancients discovered that the planets revolved in recurring periods, and thus connected the observations of their motions according to the Idea of *Time*. Kepler discovered that they revolved in ellipses, and thus connected the observations according to the Idea of *Space*. Newton discovered that they revolved in virtue of the Sun's attraction, and thus connected the motions according to the Idea of *Force*. The first and third of these discoveries are recognized on all hands as processes of Induction. Why is the second to be called by a different name? or what but confusion and perplexity can arise from refusing to class it with the other two? It is, you say, Description. But such Description is a kind of Induction, and must be spoken of as Induction, if we are to speak of Induction as the process by which Science is formed: for the three steps are all, the second in the same sense as the first and third, in co-ordination with them, steps in the formation of astronomical science.

18. But, says Mr. Mill (i. 363), "it is a fact surely that the planet does describe an ellipse, and a fact which we could see if we had adequate visual organs and a suitable position." To this I should reply: "Let it be so; and it is a fact, surely, that the planet does move period-

ically: it is a fact, surely, that the planet is attracted by the sun. Still, therefore, the asserted distinction fails to find a ground." Perhaps Mr. Mill would remind us that the elliptical form of the orbit is a fact which we could see if we had adequate visual organs and a suitable position: but that force is a thing which we cannot see. But this distinction also wil not bear handling. Can we not see a tree blown down by a storm, or a rock blown up by gunpowder? Do we not here see force:—see it, that is, by its effects, the only way in which we need to see it in the case of a planet, for the purposes of our argument? Are not such operations of force, Facts which may be the objects of sense? and is not the operation of the sun's Force a Fact of the same kind, just as much as the elliptical form of orbit which results from the action? If the latter be "surely a Fact," the former is a Fact no less surely.

19. In truth, as I have repeatedly had occasion to remark, all attempts to frame an argument by the exclusive or emphatic appropriation of the term *Fact* to particular cases, are necessarily illusory and inconclusive. There is no definite and stable distinction between Facts and Theories; Facts and Laws; Facts and Inductions. Inductions, Laws, Theories, which are true, *are* Facts. Facts involve Inductions. It is a fact that the moon is attracted by the earth, just as much as it is a Fact that an apple falls from a tree. That the former fact is collected by a more distinct and conscious Induction, does not make it the less a Fact. That the orbit of Mars is a Fact—a true Description of the path—does not make it the less a case of Induction.

20. There is another argument which Mr. Mill employs in order to show that there is a difference between mere colligation which is description, and induction in the more proper sense of the term. He notices with commendation a remark which I had made (i. 364), that at different stages of the progress of science the facts had been successfully connected by means of very different conceptions, while yet the later conceptions have not contradicted, but included, so far as they were true, the earlier: thus the ancient Greek representation of the motions of the planets by means of epicycles and eccentrics, was to a certain degree of accuracy true, and is not negatived, though superseded, by the modern representation of the planets as describing ellipses round the sun. And he then reasons that this, which is thus true of Descriptions, cannot be true of Inductions. He says (i. 367), "Dif-

ferent descriptions therefore may be all true: but surely not different explanations." He then notices the various explanations of the motions of the planets—the ancient doctrine that they are moved by an inherent virtue; the Cartesian doctrine that they are moved by impulse and by vortices; the Newtonian doctrine that they are governed by a central force; and he adds, "Can it be said of these, as was said of the different descriptions, that they are all true as far as they go? Is it not true that one only can be true in any degree, and that the other two must be altogether false?"

21. And to this questioning, the history of science compels me to reply very distinctly and positively, in the way which Mr. Mill appears to think extravagant and absurd. I am obliged to say, Undoubtedly, all these explanations *may* be true and consistent with each other, and would be so if each had been followed out so as to show in what manner it could be made consistent with the facts. And this was, in reality, in a great measure done.[2] The doctrine that the heavenly bodies were moved by vortices was successively modified, so that it came to coincide in its results with the doctrine of an inverse-quadratic centripetal force, as I have remarked in the *History*.[3] When this point was reached, the vortex was merely a machinery, well or ill devised, for producing such a centripetal force, and therefore did not contradict the doctrine of a centripetal force. Newton himself does not appear to have been averse to explaining gravity by impulse. So little is it true that if the one theory be true the other must be false. The attempt to explain gravity by the impulse of streams of particles flowing through the universe in all directions, which I have mentioned in the *Philosophy*,[4] is so far from being inconsistent with the Newtonian theory, that it is founded entirely upon it. And even with regard to the doctrine, that the heavenly bodies move by an inherent virtue; if this doctrine had been maintained in any such way that it was brought to agree with the facts, the inherent virtue must have had its laws determined; and then, it would have been found that the virtue had a reference to the central body; and so, the "inherent virtue" must have coincided in its effect with the Newtonian force; and then, the two explanations would agree, except so far as the word "inherent" was concerned. And if such a part of an earlier theory as this word *inherent* indicates, is found to be untenable, it is of course rejected in the transition to later and more exact theories, in Inductions of this kind, as well as in what Mr. Mill

calls Descriptions. There is therefore still no validity discoverable in the distinction which Mr. Mill attempts to draw between "descriptions" like Kepler's law of elliptical orbits, and other examples of induction.

22. When Mr. Mill goes on to compare what he calls different predictions—the first, the true explanation of eclipses by the shadows which the planets and satellites cast upon one another, and the other, the belief that they will occur whenever some great calamity is impending over mankind, I must reply, as I have stated already, (Art. 17), that to class such superstitions as the last with cases of Induction, appears to me to confound all use of words, and to prevent, as far as it goes, all profitable exercises of thought. What possible advantage can result from comparing (as if they were alike) the relation of two descriptions of a phenomenon, each to a certain extent true, and therefore both consistent, with the relation of a scientific truth to a false and baseless superstition?

23. But I may make another remark on this example, so strangely introduced. If, under the influence of fear and superstition, men may make such mistakes with regard to laws of nature, as to imagine that eclipses portend calamities, are they quite secure from mistakes in *description*? Do not the very persons who tell us how eclipses predict disasters, also describe to us fiery swords seen in the air, and armies fighting in the sky? So that even in this extreme case, at the very limit of the rational exercise of human powers, there is nothing to distinguish Description from Induction.

I shall now leave the reader to judge whether this feature in the history of science,—that several views which appear at first quite different are yet all true,—which Mr. Mill calls a curious and interesting remark of mine, and which he allows to be "strikingly true" of the Inductions which he calls *Descriptions,* (i. 364) is, as he says, "unequivocally false" of other Inductions. And I shall confide in having general assent with me, when I continue to speak of Kepler's *Induction* of the elliptical orbits.

I now proceed to another remark.

#### *III. In Discovery a new Conception is introduced*

24. There is a difference between Mr. Mill and me in our view of the essential elements of this Induction of Kepler, which affects all

other cases of Induction, and which is, I think, the most extensive and important of the differences between us. I must therefore venture to dwell upon it a little in detail.

I conceive that Kepler, in discovering the law of Mars's motion, and in asserting that the planet moved in an ellipse, did this;—he bound together particular observations of separate places of Mars by the notion, or, as I have called it, the *conception,* of an *ellipse,* which was supplied by his own mind. Other persons, and he too, before he made this discovery, had present to their minds the facts of such separate successive positions of the planet; but could not bind them together rightly, because they did not apply to them this conception of an *ellipse.* To supply this conception, required a special preparation, and a special activity in the mind of the discoverer. He, and others before him, tried other ways of connecting the special facts, none of which fully succeeded. To discover such a connexion, the mind must be conversant with certain relations of space, and with certain kinds of figures. To discover the right figure was a matter requiring research, invention, resource. To hit upon the right conception is a difficult step; and when this step is once made, the facts assume a different aspect from what they had before: that done, they are seen in a new point of view; and the catching this point of view, is a special mental operation, requiring special endowments and habits of thought. Before this, the facts are seen as detached, separate, lawless; afterwards, they are seen as connected, simple, regular; as parts of one general fact, and thereby possessing innumerable new relations before unseen. Kepler, then, I say, bound together the facts by superinducing upon them the *conception* of an *ellipse;* and this was an essential element in his Induction.

25. And there is the same essential element in all Inductive discoveries. In all cases, facts, before detached and lawless, are bound together by a new thought. They are reduced to law, by being seen in a new point of view. To catch this new point of view, is an act of the mind, springing from its previous preparation and habits. The facts, in other discoveries, are brought together according to other relations, or, as I have called them, *Ideas;*—the Ideas of Time, of Force, of Number, of Resemblance, of Elementary Composition, of Polarity, and the like. But in all cases, the mind performs the operation by an apprehension of some such relations; by singling out the one true relation; by

combining the apprehension of the true relation with the facts; by applying to them the Conception of such a relation.

26. In previous writings, I have not only stated this view generally, but I have followed it into detail, exemplifying it in the greater part of the History of the principal Inductive Sciences in succession. I have pointed out what are the Conceptions which have been introduced in every prominent discovery in those sciences; and have noted to which of the above Ideas, or of the like Ideas, each belongs. The performance of this task is the office of the greater part of my *Philosophy of the Inductive Sciences.* For that work is, in reality, no less historical than the *History* which preceded it. The *History of the Inductive Sciences* is the history of the discoveries, mainly so far as concerns the *Facts* which were brought together to form sciences. The *Philosophy* is, in the first ten Books, the history of the *Ideas* and *Conceptions,* by means of which the facts were connected, so as to give rise to scientific truths. It would be easy for me to give a long list of the Ideas and Conceptions thus brought into view, but I may refer any reader who wishes to see such a list, to the Tables of Contents of the *History,* and of the first ten Books of the *Philosophy.*

27. That these Ideas and Conceptions are really distinct elements of the scientific truths thus obtained, I conceive to be proved beyond doubt, not only by considering that the discoveries never were made, nor could be made, till the right Conception was obtained, and by seeing how difficult it often was to obtain this element; but also, by seeing that the Idea and the Conception itself, as distinct from the Facts, was, in almost every science, the subject of long and obstinate controversies;—controversies which turned upon the possible relations of Ideas, much more than upon the actual relations of Facts. The first ten Books of the *Philosophy* to which I have referred, contain the history of a great number of these controversies. These controversies make up a large portion of the history of each science; a portion quite as important as the study of the facts; and a portion, at every stage of the science, quite as essential to the progress of truth. Men, in seeking and obtaining scientific knowledge, have always shown that they found the formation of right conceptions in their own minds to be an essential part of the process.

28. Moreover, the presence of a Conception of the mind as a spe-

cial element of the inductive process, and as the tie by which the particular facts are bound together, is further indicated, by there being some special new *term* or *phrase* introduced in every induction; or at least some term or phrase thenceforth steadily applied to the facts, which had not been applied to them before; as when Kepler asserted that Mars moved round the sun in an *elliptical orbit,* or when Newton asserted that the planets *gravitate* towards the sun; these new terms, *elliptical orbit,* and *gravitate,* mark the new conceptions on which the inductions depend. I have in the *Philosophy*[5] further illustrated this application of "technical terms," that is, fixed and settled terms, in every inductive discovery; and have spoken of their use in enabling men to proceed from each such discovery to other discoveries more general. But I notice these terms here, for the purpose of showing the existence of a conception in the discoverer's mind, corresponding to the term thus introduced; which conception, the term is intended to convey to the minds of those to whom the discovery is communicated.

29. But this element of discovery,—right conceptions supplied by the mind in order to bind the facts together,—Mr. Mill denies to be an element at all. He says, of Kepler's discovery of the elliptical orbit (i. 363), "It superadded nothing to the particular facts which it served to bind together;" yet he adds, "except indeed the knowledge that a resemblance existed between the planetary orbit and other ellipses;" that is, except the knowledge that it *was* an ellipse;—precisely the circumstance in which the discovery consisted. Kepler, he says, "asserted as a fact that the planet moved in an ellipse. But this fact, which Kepler did not add to, but found in the motion of the planet . . . was the very fact, the separate parts of which had been separately observed; it was the sum of the different observations."

30. That the fact of the elliptical motion was not merely the *sum* of the different observations, is plain from this, that other persons, and Kepler himself before his discovery, did not find it by adding together the observations. The fact of the elliptical orbit was not the sum of the observations *merely;* it was the sum of the observations, *seen under a new point of view,* which point of view Kepler's mind supplied. Kepler found it in the facts, because it was there, no doubt, for one reason; but also, for another, because he had, in his mind, those relations of thought which enabled him to find it. We may illustrate this by a fa-

miliar analogy. We too find the law in Kepler's book; but if we did not understand Latin, we should not find it there. We must learn Latin in order to find the law in the book. In like manner, a discoverer must know the language of science, as well as look at the book of nature, in order to find scientific truth. All the discussions and controversies respecting Ideas and Conceptions of which I have spoken, may be looked upon as discussions and controversies respecting the grammar of the language in which nature speaks to the scientific mind. Man is the *Interpreter* of Nature; not the Spectator merely, but the Interpreter. The study of the language, as well as the mere sight of the characters, is requisite in order that we may read the inscriptions which are written on the face of the world. And this study of the language of nature, that is, of the necessary coherencies and derivations of the relations of phenomena, is to be pursued by examining Ideas, as well as mere phenomena;—by tracing the formation of Conceptions, as well as the accumulation of Facts. And this is what I have tried to do in the books already referred to.

31. Mr. Mill has not noticed, in any considerable degree, what I have said of the formation of the Conceptions which enter into the various sciences; but he has, in general terms, denied that the Conception is anything different from the facts themselves. "If," he says (i. 301), "the facts are rightly classed under the conceptions, it is because there is in the facts themselves, something of which the conception is a copy." But it is a copy which cannot be made by a person without peculiar endowments; just as a person cannot copy an ill-written inscription, so as to make it convey sense, unless he understand the language. "Conceptions," Mr. Mill says (ii. 217), "do not develop themselves from within, but are impressed from without." But what comes from without is not enough: they must have both origins, or they cannot make knowledge. "The conception," he says again (ii. 221), "is not furnished *by* the mind till it has been furnished *to* the mind." But it is furnished to the mind by its own activity, operating according to its own laws. No doubt, the conception may be formed, and in cases of discovery, must be formed, by the suggestion and excitement which the facts themselves produce; and must be so moulded as to agree with the facts. But this does not make it superfluous to examine, out of what *materials* such conceptions are formed,

and *how* they are capable of being moulded so as to express laws of nature; especially, when we see how large a share this part of discovery—the examination how our ideas can be modified so as to agree with nature,—holds, in the history of science.

32. I have already (Art. 28) given, as evidence that the conception enters as an element in every induction, the constant introduction in such cases, of a new fixed term or phrase. Mr. Mill (ii. 282) notices this introduction of a new phrase in such cases as important, though he does not appear willing to allow that it is necessary. Yet the necessity of the conception at least, appears to result from the considerations which he puts forward. "What darkness," he says, "would have been spread over geometrical demonstration, if wherever the word *circle* is used, the definition of a circle was inserted instead of it." "If we want to make a particular combination of ideas permanent in the mind, there is nothing which clenches it like a name specially devoted to express it." In my view, the new conception is the *nail* which connects the previous notions, and the name, as Mr. Mill says, *clenches* the junction.

33. I have above (Art. 30) referred to the difficulty of getting hold of the right conception, as a proof that induction is not a mere juxtaposition of facts. Mr. Mill does not dispute that it is often difficult to hit upon the right conception. He says (i. 360), "that a conception of the mind is introduced, is indeed most certain, and Mr. Whewell has rightly stated elsewhere, that to hit upon the right conception is often a far more difficult, and more meritorious achievement, than to prove its applicability when obtained. "But," he adds, "a conception implies and corresponds to something conceived; and although the conception itself is not in the facts, but in our mind, it must be a conception of something which really is in the facts." But to this I reply, that its being really in the facts, does not help us at all towards knowledge, if we cannot see it there. As the poet says,

> It is the mind that sees: the outward eyes
> Present the object, but the mind descries.

And this is true of the sight which produces knowledge, as well as of the sight which produces pleasure and pain, which is referred to in the Tale.

34. Mr. Mill puts his view, as opposed to mine, in various ways, but, as will easily be understood, the answers which I have to offer are in all cases nearly to the same effect. Thus, he says (ii. 216), "the tardy development of several of the physical sciences, for example, of Optics, Electricity, Magnetism, and the higher generalizations of Chemistry, Mr. Whewell ascribes to the fact that mankind had not yet possessed themselves of the idea of Polarity, that is, of opposite properties in opposite directions. But what was there to suggest such an idea, until by a separate examination of several of these different branches of knowledge it was shown that the facts of each of them did present, in some instances at least, the curious phenomena of opposite properties in opposite directions?" But on this I observe, that these facts did not, nor do yet, present this conception to ordinary minds. The opposition of properties, and even the opposition of directions, which are thus apprehended by profound cultivators of science, are of an abstruse and recondite kind; and to conceive any one kind of polarity in its proper generality, is a process which few persons hitherto appear to have mastered; still less, have men in general come to conceive of them all as modifications of a general notion of Polarity. The description which I have given of Polarity in general, "opposite properties in opposite directions," is of itself a very imperfect account of the manner in which corresponding antitheses are involved in the portions of science into which Polar relations enter. In excuse of its imperfection, I may say, that I believe it is the first attempt to define Polarity in general; but yet, the conception of Polarity has certainly been strongly and effectively present in the minds of many of the sagacious men who have discovered and unravelled polar phenomena. They attempted to convey this conception, each in his own subject, sometimes by various and peculiar expressions, sometimes by imaginary mechanism by which the antithetical results were produced; their mode of expressing themselves being often defective or imperfect, often containing what was superfluous; and their meaning was commonly very imperfectly apprehended by most of their hearers and readers. But still, the conception was there, gradually working itself into clearness and distinctness, and in the mean time, directing their experiments, and forming an essential element of their discoveries. So far would it be from a sufficient statement of the case to say, that they conceived

polarity because they saw it;—that they saw it as soon as it came into view;—and that they described it as they saw it.

35. The way in which such conceptions acquire clearness and distinctness is often by means of Discussions of Definitions. To define well a thought which already enters into trains of discovery, is often a difficult matter. The business of such definition is a part of the business of discovery. These, and other remarks connected with these, which I had made in the *Philosophy*, Mr. Mill has quoted and adopted (ii. 242). They appear to me to point very distinctly to the doctrine to which he refuses his assent,—that there is a special process in the mind, in addition to the mere observation of facts, which is necessary at every step in the progress of knowledge. The Conception must be *formed* before it can be *defined.* The Definition gives the last stamp of distinctness to the Conception; and enables us to express, in a compact and lucid form, the new scientific propositions into which the new Conception enters.

36. Since Mr. Mill assents to so much of what has been said in the *Philosophy,* with regard to the process of scientific discovery, how, it may be asked, would he express these doctrines so as to exclude that which he thinks erroneous? If he objects to our saying that when we obtain a new inductive truth, we connect phenomena by applying to them a new Conception which fits them, in what terms would he describe the process? If he will not agree to say, that in order to discover the law of the facts, we must find an appropriate Conception, what language would he use instead of this? This is a natural question; and the answer cannot fail to throw light on the relation in which his views and mine stand to each other.

Mr. Mill would say, I believe, that when we obtain a new inductive law of facts, we find something in which the facts *resemble each other;* and that the business of making such discoveries is the business of discovering such resemblances. Thus, he says (of me,) (ii. 211), "his Colligation of Facts by means of appropriate Conceptions, is but the ordinary process of finding by a comparison of phenomena, in what consists their agreement or resemblance." And the Methods of experimental Inquiry which he gives (i. 450, &c.), proceed upon the supposition that the business of discovery may be thus more properly described.

37. There is no doubt that when we discover a law of nature by induction, we find some point in which all the particular facts agree. All the orbits of the planets agree in being ellipses, as Kepler discovered; all falling bodies agree in being acted on by a uniform force, as Galileo discovered; all refracted rays agree in having the sines of incidence and refraction in a constant ratio, as Snell discovered; all the bodies in the universe agree in attracting each other, as Newton discovered; all chemical compounds agree in being constituted of elements in definite proportions, as Dalton discovered. But it appears to me a most scanty, vague, and incomplete account of these steps in science, to say that the authors of them discovered something in which the facts in each case agreed. The point in which the cases agree, is of the most diverse kind in the different cases—in some, a relation of space, in others, the action of a force, in others, the mode of composition of a substance;—and the point of agreement, visible to the discoverer alone, does not come even into his sight, till after the facts have been connected by thoughts of his own, and regarded in points of view in which he, by his mental acts, places them. It would seem to me not much more inappropriate to say, that an officer, who disciplines his men till they move together at the word of command, does so by finding something in which they agree. If the power of consentaneous motion did not exist in the individuals, he could not create it: but that power being there, he finds it and uses it. Of course I am aware that the parallel of the two cases is not exact; but in the one case, as in the other, that in which the particular things are found to agree, is something formed in the mind of him who brings the agreement into view.

### *IV. Mr. Mill's Four Methods of Inquiry*

38. Mr. Mill has not only thus described the business of scientific discovery; he has also given rules for it, founded on this description. It may be expected that we should bestow some attention upon the methods of inquiry which he thus proposes. I presume that they are regarded by his admirers as among the most valuable parts of his book; as certainly they cannot fail to be, if they describe methods of scientific inquiry in such a manner as to be of use to the inquirer.

Mr. Mill enjoins four methods of experimental inquiry, which he calls *the Method of Agreement, the Method of Difference, the Method of Residues,* and *the Method of Concomitant Variations.*[6] They are all described by formulæ of this kind:—Let there be, in the observed facts, combinations of antecedents, *ABC, BC, ADE,* &c. and combinations of corresponding consequents, *abc, bc, ade,* &c.; and let the object of inquiry be, the consequence of some cause *A,* or the cause of some consequence *a.* The Method of Agreement teaches us, that when we find by experiment such facts as *abc* the consequent of *ABC,* and *ade* the consequent of *ADE,* then *a* is the consequent of *A.* The Method of Difference teaches us that when we find such facts as *abc* the consequent of *ABC,* and *bc* the consequent of *BC,* then *a* is the consequent of *A.* The Method of Residues teaches us, that if *abc* be the consequent of *ABC,* and if we have already ascertained that the effect of *A* is *a,* and the effect of *B* is *b,* then we may infer that the effect of *C* is *c.* The Method of Concomitant Variations teaches us, that if a phenomenon *a* varies according as another phenomenon *A* varies, there is some connexion of causation direct or indirect, between *A* and *a.*

39. Upon these methods, the obvious thing to remark is, that they take for granted the very thing which is most difficult to discover, the reduction of the phenomena to formulæ such as are here presented to us. When we have any set of complex facts offered to us; for instance, those which were offered in the cases of discovery which I have mentioned,—the facts of the planetary paths, of falling bodies, of refracted rays, of cosmical motions, of chemical analysis; and when, in any of these cases, we would discover the law of nature which governs them, or if any one chooses so to term it, the feature in which all the cases agree, where are we to look for our *A, B, C* and *a, b, c*? Nature does not present to us the cases in this form; and how are we to reduce them to this form? You say, *when* we find the combination of *ABC* with *abc* and *ABD* with *abd,* then we may draw our inference. Granted: but when and where are we to find such combinations? Even now that the discoveries are made, who will point out to us what are the *A, B, C,* and *a, b, c* elements of the cases which have just been enumerated? Who will tell us which of the methods of inquiry those historically real and successful inquiries exemplify? Who will carry these formulæ through the history of the sciences, as they have really grown up; and show us that these four methods have been operative in their forma-

tion; or that any light is thrown upon the steps of their progress by reference to these formulæ?

40. Mr. Mill's four methods have a great resemblance to Bacon's "Prerogatives of Instances;" for example, the Method of Agreement to the *Instantiæ Ostensivæ;* the Method of Differences to the *Instantiæ Absentiæ in Proximo,* and the *Instantiæ Crucis;* the Method of Concomitant Variations to the *Instantiæ Migrantes.* And with regard to the value of such methods, I believe all study of science will convince us more and more of the wisdom of the remarks which Sir John Herschel has made upon them.[7]

> It has always appeared to us, we must confess, that the help which the classification of instances under their different titles of prerogative, affords to inductions, however just such classification may be in itself, is yet more apparent than real. The force of the instance must be felt in the mind before it can be referred to its place in the system; and before it can be either referred or appreciated it must be known; and when it *is* appreciated, we are ready enough to weave our web of induction, without greatly troubling ourselves whence it derives the weight we acknowledge it to have in our decisions. . . . No doubt such instances as these are highly instructive; but the difficulty in physics is to find such, not to perceive their force when found.

### *V. His Examples*

41. If Mr. Mill's four methods had been applied by him in his book to a large body of conspicuous and undoubted examples of discovery, well selected and well analysed, extending along the whole history of science, we should have been better able to estimate the value of these methods. Mr. Mill has certainly offered a number of examples of his methods; but I hope I may say, without offence, that they appear to me to be wanting in the conditions which I have mentioned. As I have to justify myself for rejecting Mr. Mill's criticism of doctrines which I have put forward, and examples which I have adduced, I may, I trust, be allowed to offer some critical remarks in return, bearing upon the examples which he has given, in order to illustrate his doctrines and precepts.

42. The first remark which I have to make is, that a large proportion of his examples (i. 480, &c.) is taken from one favourite author; who, however great his merit may be, is too recent a writer to have

had his discoveries confirmed by the corresponding investigations and searching criticisms of other labourers in the same field, and placed in their proper and permanent relation to established truths; these alleged discoveries being, at the same time, principally such as deal with the most complex and slippery portions of science, the laws of vital action. Thus Mr. Mill had adduced, as examples of discoveries, Prof. Liebig's doctrine—that death is produced by certain metallic poisons through their forming indecomposable compounds; that the effect of respiration upon the blood consists in the conversion of peroxide of iron into protoxide—that the antiseptic power of salt arises from its attraction for moisture—that chemical action is contagious; and others. Now supposing that we have no doubt of the truth of these discoveries, we must still observe that they cannot wisely be cited, in order to exemplify the nature of the progress of knowledge, till they have been verified by other chemists, and worked into their places in the general scheme of chemistry; especially, since it is tolerably certain that in the process of verification, they will be modified and more precisely defined. Nor can I think it judicious to take so large a proportion of our examples from a region of science in which, of all parts of our material knowledge, the conceptions both of ordinary persons, and even of men of science themselves, are most loose and obscure, and the genuine principles most contested; which is the case in physiology. It would be easy, I think, to point out the vague and indeterminate character of many of the expressions in which the above examples are propounded, as well as their doubtful position in the scale of chemical generalization; but I have said enough to show why I cannot give much weight to these, as cardinal examples of the method of discovery; and therefore I shall not examine in detail how far they support Mr. Mill's methods of inquiry.

43. Mr. Liebig supplies the first and the majority of Mr. Mill's examples in chapter IX. of his Book on Induction. The second is an example for which Mr. Mill states himself to be indebted to Mr. Alexander Bain; the law established being this, that (i. 487) electricity cannot exist in one body without the simultaneous excitement of the opposite electricity in some neighbouring body, which Mr. Mill also confirms by reference to Mr. Faraday's experiments on voltaic wires.

I confess I am quite at a loss to understand what there is in the doctrine here ascribed to Mr. Bain which was not known to the electricians who, from the time of Franklin, explained the phenomena of the Leyden vial. I may observe also that the mention of an "electrified atmosphere" implies a hypothesis long obsolete. The essential point in all those explanations was, that each electricity produced by induction the opposite electricity in neighbouring bodies, as I have tried to make apparent in the *History*.[8] Faraday has, more recently, illustrated this universal coexistence of opposite electricities with his usual felicity.

But the conjunction of this fact with voltaic phenomena, implies a non-recognition of some of the simplest doctrines of the subject. "Since," it is said (i. 488), "common or machine electricity, and voltaic electricity may be considered for the present purpose to be identical, Faraday wished to know, &c." I think Mr. Faraday would be much astonished to learn that he considered electricity in equilibrium, and electricity in the form of a voltaic current, to be, for any purpose, identical. Nor do I conceive that he would assent to the expression in the next page, that "from the nature of a voltaic charge, the two opposite currents necessary to the existence of each other are both accommodated in one wire." Mr. Faraday has, as it appears to me, studiously avoided assenting to this hypothesis.

44. The next example is the one already so copiously dwelt upon by Sir John Herschel, Dr. Wells's researches on the production of Dew. I have already said[9] that "this investigation, although it has sometimes been praised as an original discovery, was in fact only resolving the phenomenon into principles already discovered;" namely, the doctrine of a *constituent temperature* of vapour, the different conducting power of different bodies, and the like. And this agrees in substance with what Mr. Mill says (i. 497); that the discovery, when made, was corroborated by deduction from the known laws of aqueous vapour, of conduction, and the like. Dr. Wells's researches on Dew tended much in this country to draw attention to the general principle of Atmology; and we may see, in this and in other examples which Mr. Mill adduces, that the explanation of special phenomena by means of general principles, already established, has, for common minds, a greater charm, and is more complacently dwelt on, than the discovery of the general principles themselves.

45. The next example, (i. 502) is given in order to illustrate the Method of Residues, and is the discovery by M. Arago that a disk of copper affects the vibrations of the magnetic needle. But this apparently detached fact affords little instruction compared with the singularly sagacious researches by which Mr. Faraday discovered the cause of this effect to reside in the voltaic currents which the motion of the magnetic needle developed in the copper. I have spoken of this discovery in the *History*.[10] Mr. Mill however is quoting Sir John Herschel in thus illustrating the Method of Residues. He rightly gives the Perturbations of the Planets and Satellites as better examples of the method.[11]

46. In the next chapter (c. x.) Mr. Mill speaks of Plurality of causes and of the Intermixture of effects, and gives examples of such cases. He here teaches (i. 517) that chemical synthesis and analysis, (as when oxygen and hydrogen compose water, and when water is resolved into oxygen and hydrogen,) is properly *transformation;* but that because we find that the weight of the compound is equal to the sum of the weights of the elements, we take up the notion of chemical *composition.* I have endeavoured to show[12] that the maxim, that the sum of the weights of the elements is equal to the weight of the compounds, was, historically, not *proved* from experiment, but *assumed* in the reasonings upon experiments.

47. I have now made my remarks upon nearly all the examples which Mr. Mill gives of scientific inquiry, so far as they consist of knowledge which has really been obtained. I may mention, as points which appear to me to interfere with the value of Mr. Mill's references to examples, expressions which I cannot reconcile with just conceptions of scientific truth; as when he says (i. 523), "some other force which *impinges on* the first force;" and very frequently indeed, of the "tangential *force*," as coordinate with the centripetal force.

When he speaks (ii. 20, Note) of "the doctrine now universally received that the earth is a great natural magnet with two poles," he does not recognize the recent theory of Gauss, so remarkably coincident with a vast body of facts.[13] Indeed in his statement, he rejects no less the earlier views proposed by Halley, theorized by Euler, and confirmed by Hansteen, which show that we are compelled to assume at least *four* poles of terrestrial magnetism; which I had given an account of in the first edition of the *History*. There are several other cases which he puts, in which, the knowledge spoken of not having been yet acquired, he tells us how he would set about acquiring it; for instance, if the question were (i. 526) whether mercury be a cure for a given disease; or whether the brain be a voltaic pile (ii. 21); or whether the moon be inhabited (ii. 100); or whether all crows are black (ii. 124); I confess that I have no expectation of any advantage to philosophy from discussions of this kind.

48. I will add also, that I do not think any light can be thrown upon scientific methods, at present, by grouping along with such physical inquiries as I have been speaking of, speculations concerning the human mind, its qualities and operations. Thus he speaks (i. 508) of human characters, as exemplifying the effect of plurality of causes; of (i. 518) the phenomena of our mental nature, which are analogous to chemical rather than to dynamical phenomena; of (i. 518) the reason why susceptible persons are imaginative; to which I may add, the passage where he says (i. 444), "let us take as an example of a phenomenon which we have no means of fabricating artificially, a human mind." These, and other like examples occur in the part of his work in which he is speaking of scientific inquiry in general, not in the Book on the Logic of the Moral Sciences; and are, I think, examples more likely to lead us astray than to help our progress, in discovering the laws of Scientific Inquiry, in the ordinary sense of the term.

### *VI. Mr. Mill against Hypothesis*

49. I will now pass from Mr. Mill's methods, illustrated by such examples as those which I have been considering, to the views respecting the conditions of Scientific Induction to which I have been led, by such a survey as I could make, of the whole history of the principal Inductive Sciences; and especially, to those views to which Mr. Mill offers his objections.[14]

Mr. Mill thinks that I have been too favourable to the employment of hypotheses, as means of discovering scientific truth; and that I have countenanced a laxness of method, in allowing hypotheses to be established, merely in virtue of the accordance of their results with the phenomena. I believe I should be as cautious as Mr. Mill, in accepting mere hypothetical explanations of phenomena, in any case in which we had the phenomena, and their relations, placed before both of us in an equally clear light. I have not accepted the Undulatory theory of Heat, though recommended by so many coincidences and analogies.[15] But I see some grave reasons for not giving any great weight to Mr. Mill's admonitions;—reasons drawn from the language which he uses on the subject, and which appears to me inconsistent with the conditions of the cases to which he applies it. Thus, when he says (ii. 22) that the condition of a hypothesis accounting for all the known phenomena is "often fulfilled equally well by two conflicting hypotheses," I can only say that I know of no such case in the history of Science, where the phenomena are at all numerous and complicated; and that if such a case were to occur, one of the hypotheses might always be resolved into the other. When he says, that "this evidence (the agreement of the results of the hypothesis with the phenomena) cannot be of the smallest value, because we cannot have in the case of such an hypothesis the assurance that if the hypothesis be false it must lead to results at variance with the true facts," we must reply, with due submission, that we have, in the case spoken of, the most complete evidence of this; for any change in the hypothesis would make it incapable of accounting for the facts. When he says that "if we give ourselves the license of inventing the causes as well as their laws, a person of fertile imagination might devise a hundred modes of accounting for any given fact;" I reply, that the question is about accounting for a large and complex series of facts, of which the laws have been ascertained: and as a test of Mr. Mill's assertion, I would propose as a challenge to any person of fertile imagination to devise any *one* other hypothesis to account for the perturbations of the moon, or the coloured fringes of shadows, besides the hypothesis by which they have actually been explained with such curious completeness. This challenge has been repeatedly offered, but never in any degree accepted; and I entertain no apprehension that Mr. Mill's supposition will ever be verified by such a performance.

50. I see additional reason for mistrusting the precision of Mr. Mill's views of that accordance of phenomena with the results of a hypothesis, in several others of the expressions which he uses (ii. 23). He speaks of a hypothesis being a "*plausible* explanation of all or most of

the phenomena;" but the case which we have to consider is where it gives an *exact* representation of all the phenomena in which its results can be traced. He speaks of its being certain that the laws of the phenomena are "*in some measure analogous*" to those given by the hypothesis; the case to be dealt with being, that they are in every way identical. He speaks of this analogy being certain, from the fact that the hypothesis can be "for a moment *tenable;*" as if any one had recommended a hypothesis which is tenable only while a small part of the facts are considered, when it is inconsistent with others which a fuller examination of the case discloses. I have nothing to say, and have said nothing, in favour of hypotheses which are *not* tenable. He says there are many such "*harmonies* running through the laws of phenomena in other respects radically distinct;" and he gives as an instance, the laws of light and heat. I have never alleged such harmonies as grounds of theory, unless they should amount to identities; and if they should do this, I have no doubt that the most sober thinkers will suppose the causes to be of the same kind in the two harmonizing instances. If chlorine, iodine and brome, or sulphur and phosphorus, have, as Mr. Mill says, analogous properties, I should call these substances *analogous:* but I can see no temptation to frame an hypothesis that they are *identical* (which he seems to fear), so long as Chemistry proves them distinct. But any hypothesis of an analogy in the constitution of these elements (suppose, for instance, a resemblance in their atomic form or composition) would seem to me to have a fair claim to trial; and to be capable of being elevated from one degree of probability to another by the number, variety, and exactitude of the explanations of phenomena which it should furnish.

### *VII. Against Prediction of Facts*

51. These expressions of Mr. Mill have reference to a way in which hypotheses may be corroborated, in estimating the value of which, it appears that he and I differ. "It seems to be thought," he says (ii. 23), "that an hypothesis of the sort in question is entitled to a more favourable reception, if, besides accounting for the facts previously known, it has led to the anticipation and prediction of others which experience afterwards verified." And he adds, "Such predictions and their fulfilment are indeed well calculated to strike the ignorant vulgar;" but it is

strange, he says, that any considerable stress should be laid upon such a coincidence by scientific thinkers. However strange it may seem to him, there is no doubt that the most scientific thinkers, far more than the ignorant vulgar, have allowed the coincidence of results predicted by theory with fact afterwards observed, to produce the strongest effects upon their conviction; and that all the best-established theories have obtained their permanent place in general acceptance in virtue of such coincidences, more than of any other evidence. It was not the ignorant vulgar alone, who were struck by the return of Halley's comet, as an evidence of the Newtonian theory. Nor was it the ignorant vulgar, who were struck with those facts which did so much strike men of science, as curiously felicitous proofs of the undulatory theory of light,—the production of darkness by two luminous rays interfering in a special manner; the refraction of a single ray of light into a conical pencil; and other complex yet precise results, predicted by the theory and verified by experiment. It must, one would think, strike all persons in proportion to their thoughtfulness, that when Nature thus does our bidding, she acknowledges that we have learnt her true language. If we can predict new facts which we have not seen, as well as explain those which we have seen, it must be because our explanation is not a mere formula of observed facts, but a truth of a deeper kind. Mr. Mill says, "If the laws of the propagation of light agree with those of the vibrations of an elastic fluid in so many respects as is necessary to make the hypothesis a plausible explanation of all or most of the phenomena known at the time, it is nothing strange that they should accord with each other in one respect more." Nothing strange, if the theory be true; but quite unaccountable, if it be not. If I copy a long series of letters of which the last half-dozen are concealed, and if I guess those aright, as is found to be the case when they are afterwards uncovered, this must be because I have made out the import of the inscription. To say, that because I have copied all that I could see, it is nothing strange that I should guess those which I cannot see, would be absurd, without supposing such a ground for guessing. The notion that the discovery of the laws and causes of phenomena is a loose haphazard sort of guessing, which gives "plausible" explanations, accidental coincidences, casual "harmonies," laws, "in some measure analogous" to the true ones, suppositions "tenable" for a time, appears to me to be a misapprehension of the whole nature of science; as it certainly is inapplicable to the case to which it is principally applied by Mr. Mill.

52. There is another kind of evidence of theories, very closely approaching to the verification of untried predictions, and to which, apparently, Mr. Mill does not attach much importance, since he has borrowed the term by which I have described it, *Consilience*, but has applied it in a different manner (ii. 530, 563, 590). I have spoken, in the *Philosophy,*[16] of the *Consilience of Inductions,* as one of the *Tests of Hypotheses,* and have exemplified it by many instances; for example, the theory of universal gravitation, obtained by induction from the motions of the planets, was found to explain also that peculiar motion of the spheriodal earth which produces the Precession of the Equinoxes. This, I have said, was a striking and surprising coincidence which gave the theory a stamp of truth beyond the power of ingenuity to counterfeit. I may compare such occurrences to a case of interpreting an unknown character, in which two different inscriptions, deciphered by different persons, had given the same alphabet. We should, in such a case, believe with great confidence that the alphabet was the true one; and I will add, that I believe the history of science offers no example in which a theory supported by such consiliences, had been afterwards proved to be false.

53. Mr. Mill accepts (ii. 21) a rule of M. Comte's, that we may apply hypotheses, provided they are capable of being afterwards verified as facts. I have a much higher respect for Mr. Mill's opinion than for M. Comte's;[17] but I do not think that this rule will be found of any value. It appears to me to be tainted with the vice which I have already noted, of throwing the whole burthen of explanation upon the unexplained word *fact*—unexplained in any permanent and definite opposition to theory. As I have said, the Newtonian theory *is* a fact. Every true theory is a fact. Nor does the distinction become more clear by Mr. Mill's examples. "The vortices of Descartes would have been," he says, "a perfectly legitimate hypothesis, if it had been possible by any mode of explanation which we could entertain the hope of possessing, to bring the question whether such vortices exist or not, within the reach of our observing faculties." But this was possible, and was done. The free passage of comets through the spaces

in which these vortices should have been, convinced men that these vortices did not exist. In like manner Mr. Mill rejects the hypothesis of a luminiferous ether, "because it can neither be seen, heard, smelt, tasted, or touched." It is a strange complaint to make of the vehicle of light, that it cannot be heard, smelt, or tasted. Its vibrations *can* be seen. The fringes of shadows for instance, show its vibrations, just as the visible lines of waves near the shore show the undulations of the sea. Whether this can be touched, that is, whether it resists motion, is hardly yet clear. I am far from saying there are not difficulties on this point, with regard to *all* theories which suppose a *medium*. But there are no more difficulties of this kind in the undulatory theory of light, than there are in Fourier's theory of heat, which M. Comte adopts as a model of scientific investigation; or in the theory of voltaic *currents*, about which Mr. Mill appears to have no doubt; or of electric *atmospheres*, which, though generally obsolete, Mr. Mill appears to favour; for though it had been said that we *feel* such atmospheres, no one had said that they have the other attributes of matter.

***VIII. Newton's Vera Causa***

54. Mr. Mill conceives (ii. 17) that his own rule concerning hypotheses coincides with Newton's Rule, that the cause assumed must be a *vera causa*. But he allows that "Mr. Whewell . . . has had little difficulty in showing that his (Newton's) conception was neither precise nor consistent with itself." He also allows that "Mr. Whewell is clearly right in denying it to be necessary that the cause assigned should be a cause already known; else how could we ever become acquainted with new causes?" These points being agreed upon, I think that a little further consideration will lead to the conviction that Newton's Rule of philosophizing will best become a valuable guide, if we understand it as asserting that when the explanation of two or more different kinds of phenomena (as the revolutions of the planets, the fall of a stone, and the precession of the equinoxes,) lead us to *the same* cause, such a coincidence gives a reality to the cause. We have, in fact, in such a case, a Consilience of Inductions.

55. When Mr. Mill condemns me (ii. 24) (using, however, expressions of civility which I gladly acknowledge,) for having recognized no mode of Induction except that of trying hypothesis after

hypothesis until one is found which fits the phenomena, I must beg to remind the readers of our works, that Mr. Mill himself allows (i. 363) that the process of finding a conception which binds together observed facts "is tentative, that it consists of a succession of guesses, many being rejected until one at last occurs fit to be chosen." I must remind them also that I have given a Section upon the *Tests of Hypotheses,* to which I have just referred,—that I have given various methods of Induction, as the *Method of Gradation,* the *Method of Natural Classification,* the *Method of Curves,* the *Method of Means,* the *Method of Least Squares,* the *Method of Residues:* all which I have illustrated by conspicuous examples from the History of Science; besides which, I conceive that what I have said of the Ideas belonging to each science, and of the construction and explication of conceptions, will point out in each case, in what region we are to look for the Inductive Element in order to make new discoveries. I have already ventured to say, elsewhere, that the methods which I have given, are as definite and practical as any others which have been proposed, with the great additional advantage of being the methods by which all great discoveries in science have really been made.

***IX. Successive Generalizations***

56. There is one feature in the construction of science which Mr. Mill notices, but to which he does not ascribe, as I conceive, its due importance: I mean, that process by which we not only ascend from particular facts to a general law, but when this is done, ascend from the first general law to others more general; and so on, proceeding to the highest point of generalization. This character of the scientific process was first clearly pointed out by Bacon, and is one of the most noticeable instances of his philosophical sagacity. "There are," he says, "two ways, and can be only two, of seeking and finding truth. The one from sense and particulars, takes a flight to the most general axioms, and from these principles and their truth, settled once for all, invents and judges of intermediate axioms. The other method collects axioms from sense and particulars, ascending *continuously and by degrees,* so that in the end it arrives at the most general axioms:" meaning by *axioms,* laws or principles. The structure of the most complete sciences consists of several such steps,—*floors,* as Bacon calls them,

of successive generalization; and thus this structure may be exhibited as a kind of scientific pyramid. I have constructed this pyramid in the case of the science of Astronomy:[18] and I am gratified to find that the illustrious Humboldt approves of the design, and speaks of it as executed with complete success.[19] The capability of being exhibited in this form of successive generalizations, arising from particulars upward to some very general law, is the condition of all tolerably perfect sciences; and the steps of the successive generalizations are commonly the most important events in the history of the science.

57. Mr. Mill does not reject this process of generalization; but he gives it no conspicuous place, making it only one of three modes of reducing a law of causation into other laws. "There is," he says (i. 555), "the *subsumption* of one law under another; . . . the gathering up of several laws into one more general law which includes them all. He adds afterwards, that the general law is the *sum* of the partial ones (i. 557), an expression which appears to me inadequate, for reasons which I have already stated. The general law is not the mere sum of the particular laws. It is, as I have already said, their amount *in a new point of view*. A new conception is introduced; thus, Newton did not merely add together the laws of the motions of the moon and of the planets, and of the satellites, and of the earth; he looked at them altogether as the result of a universal force of mutual gravitation; and therein consisted his generalization. And the like might be pointed out in other cases.

58. I am the more led to speak of Mr. Mill as not having given due importance to this process of successive generalization, by the way in which he speaks in another place (ii. 525) of this doctrine of Bacon. He conceives Bacon "to have been radically wrong when he enunciates, as a universal rule, that induction should proceed from the lowest to the middle principles, and from those to the highest, never reversing that order, and consequently, leaving no room for the discovery of new principles by way of deduction[20] at all."

59. I conceive that the Inductive Table of Astronomy, to which I have already referred, shows that in that science,—the most complete which has yet existed,—the history of the science has gone on, as to its general movement, in accordance with the view which Bacon's sagacity enjoined. The successive generalizations, *so far as they were true,* were made by successive generations. I conceive also that the

Inductive Table of Optics shows the same thing; and this, without taking for granted the truth of the Undulatory Theory; for with regard to all the steps of the progress of the science, lower than that highest one, there is, I conceive, no controversy.

60. Also, the Science of Mechanics, although Mr. Mill more especially refers to it, as a case in which the highest generalizations (for example the Laws of Motion) were those earliest ascertained with any scientific exactness, will, I think, on a more careful examination of its history, be found remarkably to confirm Bacon's view. For, in that science, we have, in the first place, very conspicuous examples of the vice of the method pursued by the ancients in flying to the highest generalizations first; as when they made their false distinctions of the laws of *natural* and *violent* motions, and of *terrestrial* and *celestial* motions. Many erroneous laws of motion were asserted through neglect of facts or want of experiments. And when Galileo and his school had in some measure succeeded in discovering some of the true laws of the motions of terrestrial bodies, they did not at once assert them as general: for they did not at all apply those laws to the celestial motions. As I have remarked, all Kepler's speculations respecting the causes of the motions of the planets, went upon the supposition that the First Law of terrestrial Motion did not apply to celestial bodies; but that, on the contrary, some continual force was requisite to keep up, as well as to originate, the planetary motions. Nor did Descartes, though he enunciated the Laws of Motion with more generality than his predecessors, (but not with exactness,) venture to trust the planets to those laws; on the contrary, he invented his machinery of Vortices in order to keep up the motions of the heavenly bodies. Newton was the first who extended the laws of terrestrial motion to the celestial spaces; and in doing so, he used all the laws of the celestial motions which had previously been discovered by more limited inductions. To these instances, I may add the gradual generalization of the third Law of motion by Huyghens, the Bernoullis, and Herman, which I have described in the *History*[21] as preceding that Period of Deduction, to which the succeeding narrative[22] is appropriated. In Mechanics, then, we have a cardinal example of the historically gradual and successive ascent of science from particulars to the most general laws.

61. The Science of Hydrostatics may appear to offer a more fa-

vourable example of the ascent to the most general laws, without going through the intermediate particular laws; and it is true, with reference to this science, as I have observed,[23] that it does exhibit the *peculiarity* of our possessing the most general principles on which the phenomena depend, and from which many cases of special facts are explained by deduction; while other cases cannot be so explained, from the want of principles intermediate between the highest and the lowest. And I have assigned, as the reason of this peculiarity, that the general principles of the Mechanics of Fluids were not obtained with reference to the science itself, but by extension from the sister science of the Mechanics of Solids. The two sciences are parts of the same Inductive Pyramid; and having reached the summit of this Pyramid on one side, we are tempted to descend on the other from the highest generality to more narrow laws. Yet even in this science, the best part of our knowledge is mainly composed of inductive laws, obtained by inductive examination of particular classes of facts. The mere mathematical investigations of the laws of waves, for instance, have not led to any results so valuable as the experimental researches of Bremontier, Emy, the Webers, and Mr. Scott Russell. And in like manner in Acoustics, the Mechanics of Elastic Fluids,[24] the deductions of mathematicians made on general principles have not done so much for our knowledge, as the cases of vibrations of plates and pipes examined experimentally by Chladni, Savart, Mr. Wheatstone and Mr. Willis. We see therefore, even in these sciences, no reason to slight the wisdom which exhorts us to ascend from particulars to intermediate laws, rather than to hope to deduce these latter better from the more general laws obtained once for all.

62. Mr. Mill himself indeed, notwithstanding that he slights Bacon's injunction to seek knowledge by proceeding from less general to more general laws, has given a very good reason why this is commonly necessary and wise. He says (ii. 526), "Before we attempt to explain deductively, from more general laws, any new class of phenomena, it is desirable to have gone as far as is practicable in ascertaining the empirical laws of these phenomena; so as to compare the results of deduction, not with one individual instance after another, but with general propositions expressive of the points of agreement which have been found among many instances. For," he adds with great

justice, "if Newton had been obliged to verify the theory of gravitation, not by deducing from it Kepler's laws, but by deducing all the observed planetary positions which had served Kepler to establish those laws, the Newtonian theory would probably never have emerged from the state of an hypothesis." To which we may add, that it is certain, from the history of the subject, that in that case the hypothesis would never have been framed at all.

### X. *Mr. Mill's Hope from Deduction*

63. Mr. Mill expresses a hope of the efficacy of Deduction, rather than Induction, in promoting the future progress of Science; which hope, so far as the physical sciences are concerned, appears to me at variance with all the lessons of the history of those sciences. He says (i. 579), "that the advances henceforth to be expected even in physical, and still more in mental and social science, will be chiefly the result of deduction, is evident from the general considerations already adduced:" these considerations being, that the phenomena to be considered are very complex, and are the result of many known causes, of which we have to disentangle the results.

64. I cannot but take a very different view from this. I think that any one, looking at the state of physical science, will see that there are still a vast mass of cases, in which we do not at all know the causes, at least, in their full generality; and that the knowledge of new causes, and the generalization of the laws of those already known, can only be obtained by new *inductive* discoveries. Except by new Inductions, equal, in their efficacy for grouping together phenomena in new points of view, to any which have yet been performed in the history of science, how are we to solve such questions as those which, in the survey of what we already know, force themselves upon our minds? Such as, to take only a few of the most obvious examples—What is the nature of the connexion of heat and light? How does heat produce the expansion, liquefaction and vaporization of bodies? What is the nature of the connexion between the optical and the chemical properties of light? What is the relation between optical, crystalline and chemical polarity? What is the connexion between the atomic constitution and the physical qualities of bodies? What is the tenable definition of a mineral species? What is the true relation of the apparently different

types of vegetable life (monocotyledons, dicotyledons, and cryptogamous plants)? What is the relation of the various types of animal life (vertebrates, articulates, radiates, &c.)? What is the number, and what are the distinctions of the Vital Powers? What is the internal constitution of the earth? These, and many other questions of equal interest, no one, I suppose, expects to see solved by deduction from principles already known. But we can, in many of them, see good hope of progress by a large use of induction; including, of course, copious and careful experiments and observations.

65. With such questions before us, as have now been suggested, I can see nothing but a most mischievous narrowing of the field and enfeebling of the spirit of scientific exertion, in the doctrine that "Deduction is the great scientific work of the present and of future ages;" and that "A revolution is peaceably and progressively effecting itself in philosophy the reverse of that to which Bacon has attached his name." I trust, on the contrary, that we have many new laws of nature still to discover; and that our race is destined to obtain a sight of wider truths than any we yet discern, including, as cases, the general laws we now know, and obtained from these known laws as they must be, by Induction.

66. I can see, however, reasons for the comparatively greater favour with which Mr. Mill looks upon Deduction, in the views to which he has mainly directed his attention. The explanation of remarkable phenomena by known laws of Nature, has, as I have already said, a greater charm for many minds than the discovery of the laws themselves. In the case of such explanations, the problem proposed is more definite, and the solution more obviously complete. For the process of induction includes a mysterious step, by which we pass from particulars to generals, of which step the reason always seems to be inadequately rendered by any words which we can use; and this step to most minds is not demonstrative, as to few is it given to perform it on a great scale. But the process of explanation of facts by known laws is deductive, and has at every step a force like that of demonstration, producing a feeling peculiarly gratifying to the clear intellects which are most capable of following the process. We may often see instances in which this admiration for deductive skill appears in an extravagant measure; as when men compare Laplace with

Newton. Nor should I think it my business to argue against such a preference, unless it were likely to leave us too well satisfied with what we know already, to chill our hope of scientific progress, and to prevent our making any further strenuous efforts to ascend, higher than we have yet done, the mountain-chain which limits human knowledge.

67. But there is another reason which, I conceive, operates in leading Mr. Mill to look to Deduction as the principal means of future progress in knowledge, and which is a reason of considerable weight in the subjects of research which, as I conceive, he mainly has in view. In the study of our own minds and of the laws which govern the history of society, I do not think that it is very likely that we shall hereafter arrive at any wider principles than those of which we already possess some considerable knowledge; and this, for a special reason; namely, that our knowledge in such cases is not gathered by mere external observation of a collection of external facts; but acquired by attention to internal facts, our own emotions, thoughts, and springs of action; facts are connected by ties existing in our own consciousness, and not in mere observed juxtaposition, succession, or similitude. How the character, for instance, is influenced by various causes, (an example to which Mr. Mill repeatedly refers, ii. 518 &c.), is an inquiry which may perhaps be best conducted by considering what we know of the influence of education and habit, government and occupation, hope and fear, vanity and pride, and the like, upon men's characters, and by tracing the various effects of the intermixture of such influences. Yet even here, there seems to be room for the discovery of laws in the way of experimental inquiry: for instance, what share race or family has in the formation of character; a question which can hardly be solved to any purpose in any other way than by collecting and classing instances. And in the same way, many of the principles which regulate the material wealth of states, are obtained, if not exclusively, at least most clearly and securely, by induction from large surveys of facts. Still, however, I am quite ready to admit that in Mental and Social Science, we are much less likely than in Physical Science, to obtain new truths by any process which can be distinctively termed *Induction;* and that in those sciences, what may be called *Deductions* from principles of thought and action of which we are already

conscious, or to which we assent when they are felicitously picked out of our thoughts and put into words, must have a large share; and I may add, that this observation of Mr. Mill appears to me to be important, and, in its present connexion, new.

### *XI. Fundamental opposition of our doctrines*

68. I have made nearly all the remarks which I now think it of any consequence to make upon Mr. Mill's *Logic,* so far as it bears upon the doctrines contained in my *History* and *Philosophy.* And yet there remains still untouched one great question, involving probably the widest of all the differences between him and me. I mean the question whether geometrical axioms, (and, as similar in their evidence to these, *all* axioms,) be truths derived from experience, or be necessary truths in some deeper sense. This is one of the fundamental questions of philosophy; and all persons who take an interest in metaphysical discussions, know that the two opposite opinions have been maintained with great zeal in all ages of speculation. To me it appears that there are *two* distinct elements in our knowledge, Experience, without, and the Mind, within. Mr. Mill derives all our knowledge from Experience *alone.* In a question thus going to the root of all knowledge, the opposite arguments must needs cut deep on both sides. Mr. Mill cannot deny that our knowledge of geometrical axioms and the like, *seems* to be *necessary.* I cannot deny that our knowledge, axiomatic as well as other, *never is* acquired *without experience.*

69. Perhaps ordinary readers may despair of following our reasonings, when they find that they can only be made intelligible by supposing, on the one hand, a person who thinks distinctly and yet has never seen or felt any external object; and on the other hand, a person who is transferred, as Mr. Mill supposes (ii. 117), to "distant parts of the stellar regions where the phenomena may be entirely unlike those with which we are acquainted," and where even the axiom, that every effect must have a cause, does not hold good. Nor, in truth, do I think it necessary here to spend many words on this subject. Probably, for those who take an interest in this discussion, most of the arguments on each side have already been put forwards with sufficient repetition. I have, in an "Essay on the Fundamental

Antithesis of Philosophy," and in some accompanying "Remarks," printed[25] at the end of the second edition of my *Philosophy*, given my reply to what has been said on this subject, both by Mr. Mill, and by the author of a very able critique on my *History* and *Philosophy* which appeared in the *Quarterly Review* in 1841:[26] and I will not here attempt to revive the general discussion.

70. Perhaps I may be allowed to notice, that in one part of Mr. Mill's work where this subject is treated, there is the appearance of one of the parties to the controversy pronouncing judgment in his own cause. This indeed is a temptation which it is especially difficult for an author to resist, who writes a treatise upon *Fallacies*, the subject of Mr. Mill's fifth Book. In such a treatise, the writer has an easy way of disposing of adverse opinions by classing them as "Fallacies," and putting them side by side with opinions universally acknowledged to be false. In this way, Mr. Mill has dealt with several points which are still, as I conceive, matters of controversy (ii. 357, &c.).

71. But undoubtedly, Mr. Mill has given his argument against my opinions with great distinctness in another place (i. 319). In order to show that it is merely habitual association which gives to an experimental truth the character of a necessary truth, he quotes the case of the laws of motion, which were really discovered from experiment, but are now looked upon as the only conceivable laws; and especially, what he conceives as "the *reductio ad absurdum* of the theory of inconceivableness," an opinion which I had ventured to throw out, that if we could conceive the Composition of bodies distinctly, we might be able to see that it is necessary that the modes of their composition should be definite. I do not think that readers in general will see anything absurd in the opinion, that the laws of Mechanics, and even the laws of the Chemical Composition of bodies, may depend upon principles as necessary as the properties of space and number; and that this necessity, though not at all perceived by persons who have only the ordinary obscure and confused notions on such subjects, may be evident to a mind which has, by effort and discipline, rendered its ideas of Mechanical Causation, Elementary Composition and Difference of Kind, clear and precise. It may easily be, I conceive, that while such necessary principles are perceived

to be necessary only by a few minds of highly cultivated insight, such principles as the axioms of Geometry and Arithmetic may be perceived to be necessary by *all* minds which have any habit of abstract thought at all: and I conceive also, that though these axioms are brought into distinct view by a certain degree of intellectual cultivation, they may still be much better described as conditions of experience, than as results of experience:—as laws of the mind and of its activity, rather than as facts impressed upon a mind merely passive.

### *XII. Absurdities in Mr. Mill's Logic*

72. I will not pursue the subject further: only, as the question has arisen respecting the absurdities to which each of the opposite doctrines leads, I will point out opinions connected with this subject, which Mr. Mill has stated in various parts of his book.

He holds (i. 317) that it is merely from habit that we are unable to conceive the *last point* of space or the *last instant* of time. He holds (ii. 360) that it is strange that any one should rely upon the *a priori* evidence that space or extension is infinite, or that nothing can be made of nothing. He holds (i. 304) that the first law of *motion* is *rigorously true,* but that the axioms respecting the *lever* are only *approximately* true. He holds (ii. 110) that there may be sidereal firmaments in which events succeed each other at random, without obeying any laws of causation; although one might suppose that even if space and cause are both to have their limits, still they might terminate together: and then, even on this bold supposition, we should no *where* have a world in which events were *casual.* He holds (ii. 111) that the axiom, that every event must have a cause, is established by means of an "induction by simple enumeration:" and in like manner, that the principles of number and of geometry are proved by this method of simple enumeration alone. He ascribes the proof (i. 162) of the axiom, "things which are equal to the same are equal to each other," to the fact that this proposition has been perpetually *found* true and never false. He holds (i. 338) that "In all propositions concerning numbers, a condition is implied, without which none of them would be true; and that condition is an assumption which *may be false. The condition is that* 1 = 1."

73. Mr. Mill further holds (i. 309), that it is a characteristic property of geometrical forms, that they are capable of being painted in the imagination with a distinctness equal to reality:—that our ideas of forms exactly resemble our sensations: which, it is implied, is not the case with regard to any other class of our ideas;—that we thus may have mental pictures of all possible combinations of lines and angles, which are as fit subjects of geometrical experimentation as the realities themselves. He says, that "we know that the imaginary lines exactly resemble real ones;" and that we obtain this knowledge respecting the characteristic property of the idea of space by experience; though it does not appear *how* we can compare our *ideas* with the *realities*, since we know the realities only *by* our ideas; or why this property of their resemblance should be confined to *one class* of ideas alone.

74. I have now made such remarks as appear to me to be necessary, on the most important parts of Mr. Mill's criticism of my *Philosophy*. I hope I have avoided urging any thing in a contentious manner; as I have certainly written with no desire of controversy, but only with a view to offer to those who may be willing to receive it, some explanation of portions of my previous writings. I have already said, that if this had not been my especial object, I could with pleasure have noted the passages of Mr. Mill's *Logic* which I admire, rather than the points in which we differ. I will in a very few words refer to some of these points, as the most agreeable way of taking leave of the dispute.

I say then that Mr. Mill appears to me especially instructive in his discussion of the nature of the proof which is conveyed by the syllogism; and that his doctrine, that the force of the syllogism consists in an *inductive assertion, with an interpretation added to it,* solves very happily the difficulties which baffle the other theories of this subject. I think that this doctrine of his is made still more instructive, by his excepting from it the cases of Scriptural Theology and of Positive Law (i. 260), as cases in which general propositions, not particular facts, are our original data. I consider also that the recognition of *Kinds* (i. 166) as classes in which we have, not a finite but an *inexhaustible* body of resemblances among individuals, and as groups made by nature, not by mere definition, is very valuable, as stopping the inroad to an endless train of false philosophy. I conceive that he takes the right ground in his answer to Hume's argument against miracles (ii. 183): and I admire the acuteness with which he has criticized Laplace's tenets on the Doctrine of Chances, and the candour with which he has, in the second edition, acknowledged oversights on this subject made in the first. I think that much, I may almost say all, which he says on the subject of Language, is very philosophical; for instance, what he says (ii. 238) of the way in which words acquire their meaning in common use. I especially admire the acuteness and force with which he has shown (ii. 255) how moral principles expressed in words degenerate into formulas, and yet how the formula cannot be rejected without a moral loss. This "perpetual oscillation in spiritual truths," as he happily terms it, has never, I think, been noted in the same broad manner, and is a subject of most instructive contemplation. And though I have myself refrained from associating moral and political with physical science in my study of the subject, I see a great deal which is full of promise for the future progress of moral and political knowledge in Mr. Mill's sixth Book, "On the Logic of the Moral and Political Sciences." Even his arrangement of the various methods which have been or may be followed in "the Social Science,"—"the Chemical or Experimental Method," "the Geometrical or Abstract Method," "the Physical or Concrete Deductive Method," "the Inverse Deductive or Historical Method," though in some degree fanciful and forced, abounds with valuable suggestions; and his estimate of "the interesting philosophy of the Bentham school," the main example of "the geometrical method," is interesting and philosophical. On some future occasion, I may, perhaps, venture into the region of which Mr. Mill has thus essayed to map the highways: for it is from no despair either of the great progress to be made in such truth as that here referred to, or of the effect of philosophical method in arriving at such truth, that I have, in what I have now written, confined myself to the less captivating but more definite part of the subject.

Chapter V

# Whewell on Other Theories of Scientific Method

# CRITICISM OF ARISTOTLE'S ACCOUNT OF INDUCTION*

The Cambridge Philosophical Society has willingly admitted among its proceedings not only contributions to science, but also to the philosophy of science; and it is to be presumed that this willingness will not be less if the speculations concerning the philosophy of science which are offered to the Society involve a reference to ancient authors. Induction, the process by which general truths are collected from particular examples, is one main point in such philosophy: and the comparison of the views of Induction entertained by ancient and modern writers has already attracted much notice. I do not intend now to go into this subject at any length; but there is a cardinal passage on the subject in Aristotle's *Analytics,* (*Analyt. Prior.* II.25) which I wish to explain and discuss. I will first translate it, making such emendations as are requisite to render it intelligible and consistent, of which I shall afterwards give an account.

I will number the sentences of this chapter of Aristotle in order that I may afterwards be able to refer to them readily.

§1. "We must now proceed to observe that we have to examine not only syllogisms according to the aforesaid *figures,*—syllogisms logical and demonstrative,—but also rhetorical syllogisms,—and speaking generally, any kind of proof by which belief is influenced, following any method.

§2. "All belief arises either from Syllogism or from Induction: [we must now therefore treat of Induction.]

§3. "Induction, and the Inductive Syllogism, is when by means of one extreme term we infer the other extreme term to be true of the middle term.

* From "Criticism of Aristotle's Account of Induction," *Transactions of the Cambridge Philosophical Society,* IX, Pt. I (1850), 63–72.

§4. "Thus if *A, C,* be the extremes, and *B* the mean, we have to show, by means of *C,* that *A* is true of *B.*

§5. "Thus let *A* be *long-lived; B, that which has no gall-bladder;* and *C,* particular long-lived animals, as *elephant, horse, mule.*

§6. "Then every *C* is *A,* for all the animals above named are long-lived.

§7. "Also every *C* is *B,* for all those animals are destitute of gall-bladder.

§8. "If then *B* and *C* are convertible, and the mean (*B*) does not extend further than extreme (*C*), it necessarily follows that every *B* is *A.*

§9. "For it was shown before, that, if any two things be true of the same, and if either of them be convertible with the extreme, the other of the things predicated is true of the convertible (extreme).

§10. "But we must conceive that *C* consists of a collection of all the particular cases; for Induction is applied to all the cases.

§11. "But such a syllogism is an inference of a first truth and immediate proposition.

§12. "For when there is a mean term, there is a demonstrative syllogism through the mean; but when there is not a mean, there is proof by Induction.

§13. "And in a certain way, Induction is contrary to Syllogism; for Syllogism proves, by the middle term, that the extreme is true of the third thing: but Induction proves, by means of the third thing, that the extreme is true of the mean.

§14. "And Syllogism concluding by means of a middle term is prior by nature and more usual to us; but the proof by Induction, is more luminous."

I think that the chapter, thus interpreted, is quite coherent and intelligible; although at first there seems to be some confusion, from the author sometimes saying that Induction is a kind of Syllogism, and at other times that it is not. The amount of the doctrine is this.

When we collect a general proposition by Induction from particular cases, as for instance, that all animals destitute of gall-bladder (*acholous*), are long-lived, (if this proposition were true, of which hereafter,) we may express the process in the form of a Syllogism, if we will agree to make a collection of particular cases our middle term, and assume that the proposition in which the second extreme term occurs is convertible. Thus the known propositions are

Elephant, horse, mule, &c., are long-lived.

Elephant, horse, mule, &c., are *acholous.*

But if we suppose that the latter proposition is convertible, we shall have these propositions:

Elephant, horse, mule, &c., are long-lived.

All acholous animals are elephant, horse, mule, &c., from whence we infer, quite rigorously as to *form,*

All acholous animals are long-lived.

This mode of putting the Inductive inference shows both the strong and the weak point of the illustration of Induction by means of Syllogism. The strong point is this, that we make the inference perfect as to form, by including an indefinite collection of particular cases, elephant, horse, mule, &c., in a single term, *C*. The Syllogism then is

All *C* are long-lived.

All acholous animals are *C*.

Therefore all acholous animals are long-lived.

The weak point of this illustration is, that, at least in some instances, when the number of actual cases is necessarily indefinite, the representation of them as a single thing involves an unauthorized step. In order to give the reasoning which really passes in the mind, we must say

Elephant, horse, &c., are long-lived.

All acholous animals are *as* elephant, horse, &c.,

Therefore all acholous animals are long-lived.

This "*as*" must be introduced in order that the "all *C*" of the first proposition may be justified by the "*C*" of the second.

This step is, I say, necessarily unauthorized, where the number of particular cases is indefinite; as in the instance before us, the species of acholous animals. We do not know how many such species there are, yet we wish to be able to assert that *all* acholous animals are long-lived. In the proof of such a proposition, put in a syllogistic form, there must necessarily be a logical defect; and the above discussion shows that this defect is the substitution of the proposition,

"All acholous animals are *as* elephant, &c.," for the converse of the experimentally proved proposition, "elephant, &c., are acholous."

In instances in which the number of particular cases is limited, the necessary existence of a logical flaw in the syllogistic translation of the process is not so evident. But in truth, such a flaw exists in all cases of Induction *proper:* (for Induction by *mere enumeration* can hardly be called *Induction*). I will, however, consider for a moment the instance of a celebrated proposition which has often been taken as an example of Induction, and in which the number of particular cases is, or at least is at present supposed to be, limited. Kepler's laws, for instance the law that the planets describe ellipses, may be regarded as examples of Induction. The law was inferred, we will suppose, from an examination of the orbits of Mars, Earth, Venus. And the syllogistic illustration which Aristotle gives, will, with the necessary addition to it, stand thus,

Mars, Earth, Venus describe ellipses.

Mars, Earth, Venus are planets.

Assuming the convertibility of this last proposition, *and its universality,* (which is the necessary addition in order to make Aristotle's syllogism valid) we say

All the planets are as Mars, Earth, Venus.

Whence it follows that all the planets describe ellipses.

If, instead of this assumed universality, the astronomer had made a real enumeration, and had established the fact of each particular, he would be able to say

Saturn, Jupiter, Mars, Earth, Venus, Mercury,
describe ellipses.

Saturn, Jupiter, Mars, Earth, Venus, Mercury
are all the planets.

And he would obviously be entitled to convert the second proposition, and then to conclude that

All the planets describe ellipses.

But then, if this were given as an illustration of Induction by means of syllogism, we should have to remark, in the first place, that the conclusion that "all the planets describe ellipses," adds nothing to the major proposition, that "S., J., M., E., V., m., do so." It is merely the same proposition expressed in other words, so long as S., J., M.,

E., V., m., are supposed to be all the planets. And in the next place we have to make a remark which is more important; that the minor, in such an example, must generally be either a very precarious truth, or, as appears in this case, a transitory error. For that the planets known at any time are *all* the planets, must always be a doubtful assertion—liable to be overthrown to-night by an astronomical observation. And the assertion, as received in Kepler's time, has been overthrown. For Saturn, Jupiter, Mars, Earth, Venus, Mercury, are not all the planets. Not only have several new ones been discovered at intervals, as Uranus, Ceres, Juno, Pallas, Vesta, but we have new ones discovered every day; and any conclusion depending upon this premiss that *A, B, C, D, E, F, G, H,* to Z are all the planets, is likely to be falsified in a few years by the discovery of *A′, B′, C′*, &c. If, therefore, this were the syllogistic analysis of Induction, Kepler's discovery rested upon a false proposition; and even if the analysis were now made conformable to our present knowledge, that induction, analysed, as above, would still involve a proposition which to-morrow may show to be false. But yet no one, I suppose, doubts that Kepler's discovery was really a discovery—the establishment of a scientific truth on solid grounds; or, that it is a scientific truth for us, notwithstanding that we are constantly discovering new planets. Therefore the syllogistic analysis of it now discussed (namely, that which introduces simple enumeration as a step) is not the right analysis, and does not represent the grounds of the Inductive Truth, that all the planets describe ellipses.

It may be said that all the planets discovered since Kepler's time conform to his law, and thus confirm his discovery. This we grant: but they only *confirm* the discovery, they do not make it; they are not its groundwork. It was a discovery before these new cases were known; it was an inductive truth without them. Still, an objector might urge, if any one of these new planets had contradicted the law, it would have overturned the discovery. But this is too boldly said. A discovery which is so precise, so complex (in the phenomena which it explains), so supported by innumerable observations extending through space and time, is not so easily overturned. If we find that Uranus, or that Encke's comet, deviates from Kepler's and Newton's laws, we do not infer that these laws must be false; we say that there must be some disturbing cause in these cases. We seek, and we find these disturbing causes: in the case of Uranus, a new planet; in the case of Encke's comet, a resisting medium. Even in this case therefore, though the number of particulars is limited, the Induction was not made by a simple enumeration of all the particulars. It was made from a few cases, and when the law was discerned to be true in these, it was extended to all; the conversion and assumed universality of the proposition that "these are planets," giving us the proposition which we need for the syllogistic exhibition of Induction, "all the planets are as these."

I venture to say further, that it is plain, that Aristotle did not regard Induction as the result of simple enumeration. This is plain, in the first place, from his example. Any proposition with regard to a special class of animals, cannot be proved by simple enumeration: for the number of particular cases, that is, of animal species in the class, is indefinite at any period of zoological discovery, and must be regarded as infinite. In the next place, Aristotle says (§10 of the above extract), "We must conceive that *C* consists of a collection of all the particular cases; for induction is applied to all the cases." We must *conceive* (νοεῖν) that *C* in the major, consists of all the cases, in order that the conclusion may be true of all the cases; but we cannot *observe* all the cases. But the evident proof that Aristotle does not contemplate in this chapter an Induction by simple enumeration, is the contrast in which he places Induction and Syllogism. For Induction by simple enumeration stands in no contrast to Syllogism. The Syllogism of such Induction is quite logical and conclusive. But Induction from a comparatively small number of particular cases to a general law, does stand in opposition to Syllogism. It gives us a truth,—a truth which, as Aristotle says (§14), is more luminous than a truth proved syllogistically, though Syllogism may be *more natural and usual.* It gives us (§11) immediate propositions, obtained directly from observation, and not by a chain of reasoning: "first truths," the principles from which syllogistic reasonings may be deduced. The Syllogism proves by means of a middle term (§13) that the extreme is true of a third thing: thus, (*acholous* being the middle term):

Acholous animals are long-lived:
All elephants are acholous animals:
Therefore all elephants are long-lived.

But Induction proves by means of a third thing (namely, particular cases) that the extreme is true of the mean; thus (*acholous,* still being the middle term)

Elephants are long-lived:
Elephants are acholous animals:
Therefore acholous animals are long-lived.

It may be objected, such reasoning as this is quite inconclusive: and the answer is, that this is precisely what we, and as I believe, Aristotle, are here pointing out. Induction *is* inconclusive *as reasoning.* It is not reasoning: it is another way of getting at truth. As we have seen, no reasoning can prove such an inductive truth as this, that all planets describe ellipses. It is *known* from observation, but it is not *demonstrated.* Nevertheless, no one doubts its universal truth, (except, as aforesaid, when disturbing causes intervene). And thence, Induction is, as Aristotle says, opposed to syllogistic reasoning, and yet is a means of discovering truth: not only so, but a means of discovering primary truths, immediately derived from observation.

I have elsewhere taught that all Induction involves a *Conception* of the mind applied to facts. It may be asked whether this applies in such a case as that given by Aristotle. And I reply, that Aristotle's instance is a very instructive example of what I mean. The conception which is applied to the facts in order to make the induction possible is the want of the gall-bladder;—and Aristotle supplies us with a special term for this conception; *acholous.* But, it may be said, that the animals observed, the elephant, horse, mule, &c., are acholous, is a mere fact of observation, not a Conception. I reply that it is a *Selected* Fact, a fact selected and compared in several cases, which is what we mean by a *Conception.* That there is needed for such selection and comparison a certain activity of the mind, is evident; but this also may become more clear by dwelling a little further on the subject. Suppose that Aristotle, having a desire to know what class of animals are long-lived, had dissected for that purpose many animals; elephants, horses, cows, sheep, goats, deer and the like. How many resemblances, how many differences, must he have observed in their anatomy! He was very likely long in fixing upon any one resemblance which was common to all the long-lived. Probably he tried several other characters, before he tried the presence and absence of the gall-bladder:—perhaps, trying such characters, he found them succeed for a few cases, and then fail in others, so that he had to reject them as useless for his purpose. All the while, the absence of the gall-bladder in the long-lived animals was a fact: but it was of no use to him, because he had not selected it and drawn it forth from the mass of other facts. He was looking for a mean term to connect his first extreme, *long-lived,* with his second, the special cases. He sought this middle term in the entrails of the many animals which he used as extremes: it *was* there, but he could not find it. The fact existed, but it was of no use for the purpose of Induction, because it did not become a special Conception in his mind. He considered the animals in various points of view, it may be, as ruminant, as horned, as hoofed, and the contrary; but not as *acholous* and the contrary. When he looked at animals in that point of view,—when he took up that character as the ground of distinction, he forthwith imagined that he found a separation of long-lived and short-lived animals. When that Fact became a Conception, he obtained an inductive truth, or, at any rate, an inductive proposition.

He obtained an inductive proposition by applying the Conception *acholous* to his observation of animals. This Conception divided them into two classes; and these classes were, he fancied, long-lived and short-lived respectively. That it was the Conception, and not the Fact which enabled him to obtain his inductive proposition, is further plain from this, that the supposed Fact is not a fact. Acholous animals are not longer-lived than others. The presence or absence of the gall-bladder is no character of longevity. It is true, that in one familiar class of animals, the herbivorous kind, there is a sort of first seeming of the truth of Aristotle's asserted rule: for the horse and mule which have not the gall-bladder are longer-lived than the cow, sheep, and goat, which have it. But if we pursue the investigation further, the rule soon fails. The deer-tribe that want the gall-bladder are not longer-lived than the other ruminating animals which have it. And as a conspicuous evidence of the falsity of the rule, man and the elephant are perhaps, for their size, the longest-lived animals, and of these, man has, and the elephant has not, the organ in question. The inductive proposition, then, is false; but what we have mainly to consider is, where the fallacy enters, according to Aristotle's analysis of Induction into Syllogism. For the two premisses are still true; that elephants, &c., are long-lived; and that elephants, &c., are acholous. And it is plain that the fallacy comes in with that conversion and generalization of the latter proposition, which we have noted as necessary to Aristotle's illustration of Induction. When we say "All acholous animals are as elephants, &c.," that is, as those in their biological conditions, we say what is not true. Aristotle's condition (§8) is not complied with, that the middle term shall not extend beyond the extreme. For the character *acholous* does extend beyond the elephant and the animals biologically resembling it; it extends to deer, &c., which are not like elephants and horses, in the point in question. And thus, we see that the assumed conversion and generalization of the minor proposition, is the seat of the fallacy of false Inductions, as it is the seat of the peculiar logical character of true Inductions.

As true Inductive Propositions cannot be logically demonstrated by syllogistic rules, so they cannot be discovered by any rule. There is no formula for the discovery of inductive truth. It is caught by a peculiar sagacity, or power of divination, for which no precepts can be given. But from what has been said, we see that this sagacity shows itself in the discovery of propositions which are both *true,* and *convertible* in the sense above explained. Both these steps may be difficult. The former is often very laborious: and when the labour has been expended, and a true proposition obtained, it may turn out useless, because the proposition is not convertible. It was a matter of great labour to Kepler to prove (from calculation of observations) that Mars moves elliptically. Before he proved this, he had tried to prove many similar propositions:—that Mars moved according to the "bisection of the eccentricity,"—according to the "vicarious hypotheses,"—according to the "physical hypothesis,"—and the like; but none of these was found to be exactly true. The proposition that Mars move elliptically was proved to be true. But still, there was the question, Is it convertible? Do all the planets move as Mars moves? This was proved, (suppose,) to be true, for the Earth and Venus. But still the question remains, Do all the planets move as Mars, Earth, Venus, do? The inductive generalizing impulse boldly answers, Yes, to this question; though the rules of Syllogism do not authorize the answer, and though there remain untried cases. The inductive Philosopher tries the cases as fast as they occur, in order to confirm his previous conviction; but if he had to wait for belief and conviction till he had tried every case, he never could have belief or conviction of such a proposition at all. He is prepared to modify or add to his inductive truth according as new cases and new observations instruct him; but he does not fear that new cases or new observations will overturn an inductive proposition established by exact comparison of many complex and various phenomena.

Aristotle's example offers somewhat similar reflections. He had to establish a proposition concerning long-lived animals, which should be true, and should be susceptible of generalized conversion. To prove that the elephant, horse and mule are destitute of gall-bladder required, at least, the labour of anatomizing those animals in the seat of that organ. But this labour was not enough; for he would find those animals to agree in many other things besides in being acholous. He must have selected that character somewhat at a venture. And the guess was wrong, as a little more labour would have shown him; if for instance he had dissected deer: for they are acholous, and yet short-lived. A trial of this kind would have shown him that the extreme term, *acholous,* did extend beyond the mean, namely, animals such as elephant, horse, mule; and therefore, that the conversion was not allowable, and that the Induction was untenable. In truth, there is no relation between bile and longevity, and this example given by Aristotle of generalization from induction is an unfortunate one.

In discussing this passage of Aristotle, I have made two alterations in the text, one of which is necessary on account of the fact; the other an account of the sense. In the received text, the particular examples of long-lived animals given are *man,* horse, and mule (ἐφ' ω δὲ Γ, τὸ καθέκαστον μακρόβιον, οιον ανθρωποϛ, καὶἲπποϛ, καὶ ἡμίονοϛ). And it is afterwards said that all these are *acholous:* (ἀλλὰ καὶ τὸ β, τὸ μὴ εχον χολὴν, παντὶ υπάρχει τω Γ.) But man *has* a gall-bladder: and the fact was well known in Aristotle's time, for instance, to Hippocrates; so that it is not likely that Aristotle would have made the mistake which the text contains. But at any rate, it is a mistake; if not of the transcriber, of Aristotle; and it is impossible to reason about the passage, without correcting the mistake. The substitution of ελεφαϛ for ανθρωποϛ makes the reasoning coherent; but of course, any other acholous long-lived animal would do so equally well.

The other emendation which I have made is in §6. In the received text §6 and 7 stand thus:

6. Then every *C* is *A, for every acholous animal is long-lived* (τω δὴ Γ ολω νπὰϱχει τὸ Α, παν γὰϱ τὸ αχολον μακϱόβιον).

7. Also every *C* is *B*, for all *C* is destitute of bile.

Whence it may be inferred, says Aristotle, under certain conditions, that every *B* is *A* (τὸ *A* τω *B* νπάϱχειν) that is, that *every acholous animal is long-lived*. But this conclusion is, according to the common reading, identical with the major premiss; so that the passage is manifestly corrupt. I correct it by substituting for αχολον, Γ; and thus reading παν γὰϱ τὸ Γ μακϱόβιον "for every *C* is long-lived:" just as in the parallel sentence, 7, we have ἀλλὰ καὶ τὸ *B*, τὸ μὴ εχον χολὴν, παντὶ ὑπάϱχει τω Γ. In this way the reasoning becomes quite clear. The corrupt substitution of αχολον for Γ may have been made in various ways; which I need not suggest. As my business is with the sense of the passage, and as it makes no sense without the change, and very good sense with it, I cannot hesitate to make the emendation. And these emendations being made, Aristotle's view of the nature and force of Induction becomes, I think, perfectly clear and very instructive.

# NEWTON*

1. Bold and extensive as had been the anticipations of those whose minds were excited by the promise of the new philosophy, the discoveries of Newton respecting the mechanics of the universe, brought into view truths more general and profound than those earlier philosophers had hoped or imagined. With these vast accessions to human knowledge, men's thoughts were again set in action; and philosophers made earnest and various attempts to draw, from these extraordinary advances in science, the true moral with regard to the conduct and limits of the human understanding. They not only endeavoured to verify and illustrate, by these new portions of science, what had recently been taught concerning the methods of obtaining sound knowledge; but they were also led to speculate concerning many new and more interesting questions relating to this subject. They saw, for the first time, or at least far more clearly than before, the distinction between the inquiry into the *laws,* and into the *causes* of phenomena. They were tempted to ask, how far the discovery of causes could be carried; and whether it would soon reach, or clearly point to, the ultimate cause. They were driven to consider whether the properties which they discovered were essential properties of all matter, necessarily and primarily involved in its essence, though revealed to us at a late period by their derivative effects. These questions even now agitate the thoughts of speculative men. Some of them have already, in this work, been discussed, or arranged in the places which our view of the philosophy of these subjects assigns to them. But we must here notice them as they occurred to Newton himself and his immediate followers.

2. The general Baconian notion of the method of philosophizing,—that it consists in ascending from phenomena, through various stages of generalization, to truths of the highest order,—received, in New-

* From *Philosophy of Discovery* (1860), Ch. XVIII.

ton's discovery of the universal mutual gravitation of every particle of matter, that pointed actual exemplification, for want of which it had hitherto been almost overlooked, or at least very vaguely understood. That great truth, and the steps by which it was established, afford even now, by far the best example of the successive ascent, from one scientific truth to another,—of the repeated transition from less to more general propositions,—which we can yet produce; as may be seen in the Table which exhibits the relation of these steps in Book II. of the *Novum Organon Renovatum.* [See after page 180.] Newton himself did not fail to recognize this feature in the truths which he exhibited. Thus he says,[1] "By the way of Analysis we proceed from compounds to ingredients, as from motions to the forces producing them; and in general, from effects to their causes, and from particular causes to more general ones, till the argument ends in the most general." And in like manner in another Query:[2] "The main business of natural philosophy is to argue from phenomena without feigning hypotheses, and to deduce causes from effects, till we come to the First Cause, which is certainly not mechanical."

3. Newton appears to have had a horror of the term *hypothesis,* which probably arose from his acquaintance with the rash and illicit general assumptions of Descartes. Thus in the passage just quoted, after declaring that gravity must have some other cause than matter, he says, "Later philosophers banish the consideration of such a cause out of Natural Philosophy, feigning hypotheses for explaining all things mechanically, and referring other causes to metaphysics." In the celebrated Scholium at the end of the *Principia* he says, "Whatever is not deduced from the phenomena, is to be termed *hypothesis;* and hypotheses, whether metaphysical or physical, or occult causes, or mechanical, have no place in experimental philosophy. In this philosophy, propositions are deduced from phenomena, and rendered general by induction." And in another place, he arrests the course of his own suggestions, saying, "Verum hypotheses non fingo." I have already attempted to show that this is, in reality, a superstitious and self-destructive spirit of speculation. Some hypotheses are necessary, in order to connect the facts which are observed; some new principle of unity must be applied to the phenomena, before induction can be attempted. What is requisite is, that the hypothesis should be close

to the facts, and not connected with them by the intermediation of other arbitrary and untried facts; and that the philosopher should be ready to resign it as soon as the facts refuse to confirm it. We have seen in the *History*,[3] that it was by such a use of hypotheses, that both Newton himself, and Kepler, on whose discoveries those of Newton were based, made their discoveries. The suppositions of a force tending to the sun and varying inversely as the square of the distance; of a mutual force between all the bodies of the solar system; of the force of each body arising from the attraction of all its parts; not to mention others, also propounded by Newton,—were all hypotheses before they were verified as theories. It is related that when Newton was asked how it was that he saw into the laws of nature so much further than other men, he replied, that if it were so, it resulted from his keeping his thoughts steadily occupied upon the subject which was to be thus penetrated. But what is this occupation of thoughts, if it be not the process of keeping the phenomena clearly in view, and trying, one after another, all the plausible hypotheses which seem likely to connect them, till at last the true law is discovered? Hypotheses so used are a necessary element of discovery.

4. With regard to the details of the process of discovery, Newton has given us some of his views, which are well worthy of notice, on account of their coming from him; and which are real additions to the philosophy of this subject. He speaks repeatedly of the *analysis* and *synthesis* of observed facts; and thus marks certain steps in scientific research, very important, and not, I think, clearly pointed out by his predecessors. Thus he says,[4] "As in Mathematics, so in Natural Philosophy, the investigation of difficult things by the method of analysis ought ever to precede the method of composition. This analysis consists in making experiments and observations, and in drawing general conclusions from them by induction, and admitting of no objections against the conclusions, but such as are taken from experiments or other certain truths. And although the arguing from experiments and observations by induction be no demonstration of general conclusions; yet it is the best way of arguing which the nature of things admits of, and may be looked upon as so much the stronger, by how much the induction is more general." And he then observes, as we have quoted above, that by this way of analysis we

proceed from compounds to ingredients, from motions to forces, from effects to causes, and from less to more general causes. The *analysis* here spoken of includes the steps which in *our* Novum Organon we call the *decomposition* of facts, the exact *observation* and *measurement* of the phenomena, and the *colligation* of facts; the necessary intermediate step, the *selection* and *explication* of the appropriate conception, being passed over by Newton, in the fear of seeming to encourage the fabrication of hypotheses. The *synthesis* of which Newton here speaks consists of those steps of *deductive reasoning*, proceeding from the conception once assumed, which are requisite for the comparison of its consequences with the observed facts. This, his statement of the process of research, is, as far as it goes, perfectly exact.

5. In speaking of Newton's precepts on the subject, we are naturally led to the celebrated "Rules of Philosophizing," inserted in the second edition of the *Principia*. These rules have generally been quoted and commented on with an almost unquestioning reverence. Such Rules, coming from such an authority, cannot fail to be highly interesting to us; but at the same time, we cannot here evade the necessity of scrutinizing their truth and value, according to the principles which our survey of this subject has brought into view. The Rules stand at the beginning of that part of the *Principia* (the Third Book) in which he infers the mutual gravitation of the sun, moon, planets, and all parts of each. They are as follows:

"Rule I. We are not to admit other causes of natural things than such as both are true, and suffice for explaining their phenomena.

"Rule II. Natural effects of the same kind are to be referred to the same causes, as far as can be done.

"Rule III. The qualities of bodies which cannot be increased or diminished in intensity, and which belong to all bodies in which we can institute experiments, are to be held for qualities of all bodies whatever.

"Rule IV. In experimental philosophy, propositions collected from phenomena by induction, are to be held as true either accurately or approximately, notwithstanding contrary hypotheses; till other phenomena occur by which they may be rendered either more accurate or liable to exception."

In considering these Rules, we cannot help remarking, in the first place, that they are constructed with an intentional adaptation to the case with which Newton has to deal,—the induction of Universal Gravitation; and are intended to protect the reasonings before which they stand. Thus the first Rule is designed to strengthen the inference of gravitation from the celestial phenomena, by describing it as a *vera causa,* a true cause; the second Rule countenances the doctrine that the planetary motions are governed by mechanical forces, as terrestrial motions are; the third rule appears intended to justify the assertion of gravitation, as a *universal* quality of bodies; and the fourth contains, along with a general declaration of the authority of induction, the author's usual protest against hypotheses, levelled at the Cartesian hypotheses especially.

6. *Of the First Rule* We, however, must consider these Rules in their general application, in which point of view they have often been referred to, and have had very great authority allowed them. One of the points which has been most discussed, is that maxim which requires that the causes of phenomena which we assign should be true causes, *veræ causæ.* Of course this does not mean that they should be *the* true or right cause; for although it is the philosopher's aim to discover such causes, he would be little aided in his search of truth, by being told that it is truth which he is to seek. The rule has generally been understood to prescribe that in attempting to account for any class of phenomena, we must assume such causes only, as *from other considerations,* we know to exist. Thus gravity, which was employed in explaining the motions of the moon and planets, was already known to exist and operate at the earth's surface.

Now the Rule thus interpreted is, I conceive, an injurious limitation of the field of induction. For it forbids us to look for a cause, except among the causes with which we are already familiar. But if we follow this rule, how shall we ever become acquainted with any new cause? Or how do we know that the phenomena which we contemplate do really arise from some cause which we already truly know? If they do not, must we still insist upon making them depend upon some of our known causes; or must we abandon the study of them altogether? Must we, for example, resolve to refer the action of radiant heat to the air, rather than to any peculiar fluid or ether, because the former is known to exist, the latter is merely assumed for the purpose of explanation? But why should we do this? Why should we not endeavour to learn the cause from the effects, even if it be not already known to us? We can infer causes, which are new when we first become acquainted with them. Chemical Forces, Optical Forces, Vital Forces, are known to us only by chemical and optical and vital phenomena; must we, therefore, reject their existence or abandon their study? They do not conform to the double condition, that they shall be sufficient and *also* real: they are true, only so far as they explain the facts, but are they, therefore, unintelligible or useless? Are they not highly important and instructive subjects of speculation? And if the gravitation which rules the motions of the planets had not existed at the earth's surface;—if it had been there masked and concealed by the superior effect of magnetism, or some other extraneous force,—might not Newton still have inferred, from Kepler's laws, the tendency of the planets to the sun; and from their perturbations, their tendency to each other? His discoveries would still have been immense, if the cause which he assigned had not been a *vera causa* in the sense now contemplated.

7. But what do we mean by calling gravity a "true cause"? How do we learn its reality? Of course, by its effects, with which we are familiar;—by the weight and fall of bodies about us. These strike even the most careless observer. No one can fail to see that all bodies which we come in contact with are heavy;—that gravity acts in our neighbourhood here upon earth. Hence, it may be said, this cause is at any rate a true cause, whether it explains the celestial phenomena or not.

But if this be what is meant by a *vera causa,* it appears strange to require that in all cases we should find such a one to account for all classes of phenomena. Is it reasonable or prudent to demand that we shall reduce every set of phenomena, however minute, or abstruse, or complicated, to causes so obviously existing as to strike the most incurious, and to be familiar among men? How can we expect to find *such veræ causæ* for the delicate and recondite phenomena which an exact and skilful observer detects in chemical, or optical, or electrical experiments? The facts themselves are too fine for vulgar apprehension; their relations, their symmetries, their measures require a previous discipline to understand them. How then can their causes be found among those agencies with which the common unscientific herd of mankind are familiar? What likelihood is there that causes held for real by such persons, shall explain facts which such persons cannot see or cannot understand?

Again: if we give authority to such a rule, and require that the causes by which science explains the facts which she notes and measures and analyses, shall be causes which men, without any special study, have already come to believe in, from the effects which they casually see around them, what is this, except to make our first rude and unscientific persuasions the criterion and test of our most laborious and thoughtful inferences? What is it, but to give to ignorance and thoughtlessness the right of pronouncing upon the convictions of intense study and long-disciplined thought? "Electrical atmospheres" surrounding electrized bodies, were at one time held to be a "true cause" of the effects which such bodies produce. These atmospheres, it was said, are obvious to the senses; we feel them like a spider's web on the hands and face. Æpinus had to answer such persons, by proving that there are no atmospheres, no effluvia, but only repulsion. He thus, for a *true cause* in the vulgar sense of the term, substituted an *hypothesis;* yet who doubts that what he did was an advance in the science of electricity?

8. Perhaps some persons may be disposed to say, that Newton's Rule does not enjoin us to take those causes only which we clearly know, or suppose we know, to be really existing and operating, but only causes of *such kinds* as we have already satisfied ourselves do exist in nature. It may be urged that we are entitled to infer that the planets are governed in their motions by an attractive force, because we find, in the bodies immediately subject to observation and experiment, that such motions are produced by attractive forces, for example, by that of the earth. It may be said that we might on similar grounds infer forces which unite particles of chemical compounds, or deflect particles of light, because we see adhesion and deflection produced by forces.

But it is easy to show that the Rule, thus laxly understood, loses all significance. It prohibits no hypothesis; for all hypotheses suppose causes *such as,* in some case or other, we have seen in action. No one would think of explaining phenomena by referring them to forces and agencies altogether different from any which are known; for on this supposition, how could he pretend to reason about the effects of the assumed causes, or undertake to prove that they would explain the facts? Some close similarity with some known kind of cause is requisite, in order that the hypothesis may have the appearance of an explanation. No forces, or virtues, or sympathies, or fluids, or ethers, would be excluded by *this* interpretation of *veræ causæ.* Least of all, would such an interpretation reject the Cartesian hypothesis of vortices; which undoubtedly, as I conceive, Newton intended to condemn by his Rule. For that *such* a case as a whirling fluid, carrying bodies round a centre in orbits, does occur, is too obvious to require proof. Every eddying stream, or blast that twirls the dust in the road, exhibits examples of such action, and would justify the assumption of the vortices which carry the planets in their courses; as indeed, without doubt, such facts suggested the Cartesian explanation of the solar system. The vortices, in this mode of considering the subject, are at the least as *real* a cause of motion as gravity itself.

9. Thus the Rule which enjoins "true causes," is nugatory, if we take *veræ causæ* in the extended sense of any causes of a real *kind,* and unphilosophical, if we understand the term of *those very* causes which we familiarly suppose to exist. But it may be said that we are to designate as "true causes," not those which are collected in a loose, confused and precarious manner, by undisciplined minds, from obvious phenomena, but those which are justly and rigorously inferred. Such a cause, it may be added, gravity is; for the facts of the downward pressures and downward motions of bodies at the earth's surface lead us, by the plainest and strictest induction, to the assertion of such a force. Now to this interpretation of the Rule there is no objection; but then, it must be observed, that on this view, terrestrial gravity is inferred by the same process as celestial gravitation; and the cause is no more entitled to be called "true," because it is obtained from the former, than because it is obtained from the latter class of facts. We thus obtain an intelligible and tenable explanation of a *vera causa;* but then, by this explanation, its *verity* ceases to be distinguishable from its other condition, that it "suffices for the explanation of the phenomena." The assumption of universal gravitation ac-

counts for the fall of a stone; it also accounts for the revolutions of the Moon or of Saturn; but since both these explanations are of the same kind, we cannot with justice make the one a criterion or condition of the admissibility of the other.

10. But still, the Rule, so understood, is so far from being unmeaning or frivolous, that it expresses one of the most important tests which can be given of a sound physical theory. It is true, the explanation of one set of facts may be of the same nature as the explanation of the other class: but then, that the cause explains *both* classes, gives it a very different claim upon our attention and assent from that which it would have if it explained one class only. The very circumstance that the two explanations coincide, is a most weighty presumption in their favour. It is the testimony of two witnesses in behalf of the hypothesis; and in proportion as these two witnesses are separate and independent, the conviction produced by their agreement is more and more complete. When the explanation of two kinds of phenomena, distinct, and not apparently connected, leads us to the same cause, such a coincidence does give a reality to the cause, which it has not while it merely accounts for those appearances which suggested the supposition. This coincidence of propositions inferred from separate classes of facts, is exactly what we noticed in the *Novum Organon Renovatum* (b. ii. c. 5, sect. 3), as one of the most decisive characteristics of a true theory, under the name of *Consilience of Inductions.*

That Newton's First Rule of Philosophizing, so understood, authorizes the inferences which he himself made, is really the ground on which they are so firmly believed by philosophers. Thus when the doctrine of a gravity varying inversely as the square of the distance from the body, accounted at the same time for the relations of times and distances in the planetary orbits and for the amount of the moon's deflection from the tangent of her orbit, such a doctrine became most convincing: or again, when the doctrine of the universal gravitation of all parts of matter, which explained so admirably the inequalities of the moon's motions, also gave a satisfactory account of a phenomenon utterly different, the precession of the equinoxes. And of the same kind is the evidence in favour of the undulatory theory of light, when the assumption of the length of an undulation, to which we are led by the colours of thin plates, is found to be identical with that length which explains the phenomena of diffraction; or when the hypothesis of transverse vibrations, suggested by the facts of polarization, explains also the laws of double refraction. When such a convergence of two trains of induction points to the same spot, we can no longer suspect that we are wrong. Such an accumulation of proof really persuades us that we have to do with a *vera causa.* And if this kind of proof be multiplied;—if we again find other facts of a sort uncontemplated in framing our hypothesis, but yet clearly accounted for when we have adopted the supposition;—we are still further confirmed in our belief; and by such accumulation of proof we may be so far satisfied, as to believe without conceiving it possible to doubt. In this case, when the validity of the opinion adopted by us has been repeatedly confirmed by its sufficiency in unforeseen cases, so that all doubt is removed and forgotten, the theoretical cause takes its place among the realities of the world, and becomes a *true cause.*

11. Newton's Rule then, to avoid mistakes, might be thus expressed: That "we may, provisorily, assume such hypothetical cause as will account for any given class of natural phenomena; but that when two different classes of facts lead us to the same hypothesis, we may hold it to be a *true cause.*" And this Rule will rarely or never mislead us. There are no instances, in which a doctrine recommended in this manner has afterwards been discovered to be false. There have been hypotheses which have explained many phenomena, and kept their ground long, and have afterwards been rejected. But these have been hypotheses which explained only one class of phenomena; and their fall took place when another kind of facts was examined and brought into conflict with the former. Thus the system of eccentrics and epicycles accounted for all the observed *motions* of the planets, and was the means of expressing and transmitting all astronomical knowledge for two thousand years. But then, how was it overthrown? By considering the *distances* as well as motions of the heavenly bodies. Here was a second class of facts; and when the system was adjusted so as to agree with the one class, it was at variance with the other. These cycles and epicycles could not be true, because they could not be made a just representation of the facts. But if the measures of distance as well as of position had conspired in pointing out the cycles and epicycles, as the paths of the planets, the paths so determined could not

have been otherwise than their real paths; and the epicyclical theory would have been, at least geometrically, true.

12. *Of the Second Rule* Newton's Second Rule directs that "natural events of the *same kind* are to be referred to the *same causes,* so far as can be done." Such a precept at first appears to help us but little; for all systems, however little solid, profess to conform to such a rule. When any theorist undertakes to explain a class of facts, he assigns causes which, according to him, will by their natural action, as seen in other cases, produce the effects in question. The events which he accounts for by his hypothetical cause, are, he holds, of the same kind as those which such a cause is known to produce. Kepler, in ascribing the planetary motions to magnetism, Descartes, in explaining them by means of vortices, held that they were referring celestial motions to the causes which give rise to terrestrial motions of the same kind. The question is, *Are* the effects of the same kind? This once settled, there will be no question about the propriety of assigning them to the same cause. But the difficulty is, to determine *when* events are of the same kind. Are the motions of the planets of the same kind with the motion of a body moving freely in a curvilinear path, or do they not rather resemble the motion of a floating body swept round by a whirling current? The Newtonian and the Cartesian answered this question differently. How then can we apply this Rule with any advantage?

13. To this we reply, that there is no way of escaping this uncertainty and ambiguity, but by obtaining a clear possession of the ideas which our hypothesis involves, and by reasoning rigorously from them. Newton asserts that the planets move in free paths, acted on by certain forces. The most exact calculation gives the closest agreement of the results of this hypothesis with the facts. Descartes asserts that the planets are carried round by a fluid. The more rigorously the conceptions of force and the laws of motion are applied to this hypothesis, the more signal is its failure in reconciling the facts to one another. Without such calculation, we can come to no decision between the two hypotheses. If the Newtonian hold that the motions of the planets are *evidently* of the *same kind* as those of a body describing a curve in free space, and therefore, like that, to be explained by a force acting upon the body; the Cartesian denies that the planets do move in free

space. They are, he maintains, immersed in a plenum. It is only when it appears that comets pass through this plenum in all directions with no impediment, and that no possible form and motion of its whirlpools can explain the forces and motions which are observed in the solar system, that he is compelled to allow the Newtonian's classification of events of the *same kind.*

Thus it does not appear that this Rule of Newton can be interpreted in any distinct and positive manner, otherwise than as enjoining that, in the task of induction, we employ clear ideas, rigorous reasoning, and close and fair comparison of the results of the hypothesis with the facts. These are, no doubt, important and fundamental conditions of a just induction; but in this injunction we find no peculiar or technical criterion by which we may satisfy ourselves that we are right, or detect our errors. Still, of such general prudential rules, none can be more wise than one which thus, in the task of connecting facts by means of ideas, recommends that the ideas be clear, the facts, correct, and the chain of reasoning which connects them, without a flaw.

14. *Of the Third Rule* The Third Rule, that "qualities which are observed without exception be held to be universal," as I have already said, seems to be intended to authorize the assertion of gravitation as a universal attribute of matter. We formerly stated, in treating of Mechanical Ideas,[5] that this application of such a Rule appears to be a mode of reasoning far from conclusive. The assertion of the universality of any property of bodies must be grounded upon the reason of the case, and not upon any arbitrary maxim. Is it intended by this Rule to prohibit any further examination how far gravity is an original property of matter, and how far it may be resolved into the result of other agencies? We know perfectly well that this was not Newton's intention; since the cause of gravity was a point which he proposed to himself as a subject of inquiry. It would certainly be very unphilosophical to pretend, by this Rule of Philosophizing, to prejudge the question of such hypotheses as that of Mosotti, That gravity is the excess of the electrical attraction over electrical repulsion, and yet to adopt this hypothesis, would be to suppose electrical forces more truly universal than gravity; for according to the hypothesis, gravity, being the inequality of the attraction and repulsion, is only an accidental and

partial relation of these forces. Nor would it be allowable to urge this Rule as a reason of assuming that double stars are attracted to each other by a force varying according to the inverse square of the distance; without examining, as Herschel and others have done, the orbits which they really describe. But if the Rule is not available in such cases, what is its real value and authority? and in what cases are they exemplified?

15. In a former work,[6] it was shown that the fundamental laws of motion, and the properties of matter which these involve, are, after a full consideration of the subject, unavoidably assumed as universally true. It was further shown, that although our knowledge of these laws and properties be gathered from experience, we are strongly impelled, (some philosophers think, authorized,) to look upon these as not only universally, but necessarily true. It was also stated, that the law of gravitation, though its universality may be deemed probable, does not apparently involve the same necessity as the fundamental laws of motion. But it was pointed out that these are some of the most abstruse and difficult questions of the whole of philosophy; involving the profound, perhaps insoluble, problem of the identity or diversity of Ideas and Things. It cannot, therefore, be deemed philosophical to cut these Gordian knots by peremptory maxims, which encourage us to decide without rendering a reason. Moreover, it appears clear that the reason which is rendered for this Rule by the Newtonians is quite untenable; namely, that we know extension, hardness, and inertia, to be universal qualities of bodies by experience alone, and that we have the same evidence of experience for the universality of gravitation. We have already observed that we cannot, with any propriety, say that we *find* by experience all bodies are extended. This could not be a just assertion, unless we conceive the possibility of our finding the contrary. But who can conceive our finding by experience some bodies which are not extended? It appears, then, that the reason given for the Third Rule of Newton involves a mistake respecting the nature and authority of experience. And the Rule itself cannot be applied without attempting to decide, by the casual limits of observation, questions which necessarily depend upon the relations of ideas.

16. *Of the Fourth Rule* Newton's Fourth Rule is, that "Propositions collected from phenomena by induction, shall be held to be

true, notwithstanding contrary hypotheses; but shall be liable to be rendered more accurate, or to have their exceptions pointed out, by additional study of phenomena." This Rule contains little more than a general assertion of the authority of induction, accompanied by Newton's usual protest against hypotheses.

The really valuable part of the Fourth Rule is that which implies that a constant verification, and, if necessary, rectification, of truths discovered by induction, should go on in the scientific world. Even when the law is, or appears to be, most certainly exact and universal, it should be constantly exhibited to us afresh in the form of experience and observation. This is necessary, in order to discover exceptions and modifications if such exist: and if the law be rigorously true, the contemplation of it, as exemplified in the world of phenomena, will best give us that clear apprehension of its bearings which may lead us to see the ground of its truth.

The concluding clause of this Fourth Rule appears, at first, to imply that all inductive propositions are to be considered as merely provisional and limited, and never secure from exception. But to judge thus would be to underrate the stability and generality of scientific truths; for what man of science can suppose that we shall hereafter discover exceptions to the universal gravitation of all parts of the solar system? And it is plain that the author did not intend the restriction to be applied so rigorously; for in the Third Rule, as we have just seen, he authorizes us to infer universal properties of matter from observation, and carries the liberty of inductive inference to its full extent. The Third Rule appears to encourage us to assert a law to be universal, even in cases in which it has not been tried; the Fourth Rule seems to warn us that the law may be inaccurate, even in cases in which it has been tried. Nor is either of these suggestions erroneous; but both the universality and the rigorous accuracy of our laws are proved by reference to Ideas rather than to Experience; a truth, which, perhaps, the philosophers of Newton's time were somewhat disposed to overlook.

17. The disposition to ascribe all our knowledge to Experience, appears in Newton and the Newtonians by other indications; for instance, it is seen in their extreme dislike to the ancient expressions by which the principles and causes of phenomena were described, as the

*occult causes* of the Schoolmen, and the *forms* of the Aristotelians, which had been adopted by Bacon. Newton says,[7] that the particles of matter not only possess inertia, but also active principles, as gravity, fermentation, cohesion; he adds, "These principles I consider not as Occult Qualities, supposed to result from the Specific Forms of things, but as General Laws of Nature, by which the things themselves are formed: their truth appearing to us by phenomena, though their causes be not yet discovered. For these are manifest qualities, and their causes only are occult. And the Aristotelians gave the name of *occult qualities,* not to manifest qualities, but to such qualities only as they supposed to lie hid in bodies, and to the unknown causes of manifest effects: such as would be the causes of gravity, and of magnetick and electrick attractions, and of fermentations, if we should suppose that these forces or actions arose from qualities unknown to us, and incapable of being discovered and made manifest. Such occult qualities put a stop to the improvement of Natural Philosophy, and therefore of late years have been rejected. To tell us that every species of things is endowed with an occult specific quality by which it acts and produces manifest effects, is to tell us nothing: but to derive two or three general principles of motion from phenomena, and afterwards to tell us how the properties and actions of all corporeal things follow from these manifest principles, would be a great step in philosophy, though the causes of those principles were not yet discovered: and therefore I scruple not to propose the principles of motion above maintained, they being of very general extent, and leave their causes to be found out."

18. All that is here said is highly philosophical and valuable; but we may observe that the investigation of *specific forms* in the sense in which some writers had used the phrase, was by no means a frivolous or unmeaning object of inquiry. Bacon and others had used *form* as equivalent to *law.*[8] If we could ascertain that arrangement of the particles of a crystal from which its external crystalline form and other properties arise, this arrangement would be the *internal form* of the crystal. If the undulatory theory be true, the *form* of light is transverse vibrations: if the emission theory be maintained, the *form* of light is particles moving in straight lines, and deflected by various forces. Both the terms, *form* and *law,* imply an ideal connexion of

sensible phenomena; form supposes matter which is moulded to the form; law supposes objects which are governed by the law. The former term refers more precisely to existences, the latter to occurrences. The latter term is now the more familiar, and is, perhaps, the better metaphor: but the former also contains the essential antithesis which belongs to the subject, and might be used in expressing the same conclusions.

But occult causes, employed in the way in which Newton describes, had certainly been very prejudicial to the progress of knowledge, by stopping inquiry with a mere word. The absurdity of such pretended explanations had not escaped ridicule. The pretended physician in the comedy gives an example of an occult cause or virtue.

> Mihi demandatur
> A doctissimo Doctore
> *Quare* Opium facit dormire:
> Et ego respondeo,
> *Quia* est in eo
> *Virtus dormitiva,*
> Cujus natura est sensus assoupire.

19. But the most valuable part of the view presented to us in the quotation just given from Newton is the distinct separation, already noticed as peculiarly brought into prominence by him, of the determination of the *laws* of phenomena, and the investigation of their *causes*. The maxim, that the former inquiry must precede the latter, and that if the general laws of facts be discovered, the result is highly valuable, although the causes remain unknown, is extremely important; and had not, I think, ever been so strongly and clearly stated, till Newton both repeatedly promulgated the precept, and added to it the weight of the most striking examples.

We have seen that Newton, along with views the most just and important concerning the nature and methods of science, had something of the tendency, prevalent in his time, to suspect or reject, at least speculatively, all elements of knowledge except observation. This tendency was, however, in him so corrected and restrained by his own wonderful sagacity and mathematical habits, that it scarcely led to any opinion which we might not safely adopt. . . .

# Bibliography

I. Whewell's Major Writings in Philosophy of Science


*Astronomy and General Physics Considered with Reference to Natural Theology* (London 1833). A "Bridgewater Treatise."

"Comte and Positivism," *Macmillan's Magazine*, 13 (1866), 353–62.

"Demonstration that All Matter is Heavy," *Transactions of the Cambridge Philosophical Society*, 7 (1841), 197–207.

"Discussion of the Question: Are Cause and Effect Successive or Simultaneous?" *Transactions of the Cambridge Philosophical Society*, 7 (1842), 319–31.

*An Elementary Treatise on Mechanics*, 1st ed. (Cambridge 1819), and 5th ed. (Cambridge 1836).

"On the Fundamental Antithesis of Philosophy," *Transactions of the Cambridge Philosophical Society*, 7 (1844), 170–81.

*History of the Inductive Sciences*, 1st ed. (London 1837), 3 vols.

*The History of Scientific Ideas*, (London 1858), 2 vols. Pt. I of the 3rd ed. of *The Philosophy of the Inductive Sciences.*

*Of Induction, with Especial Reference to Mr. Mill's System of Logic* (London 1849).

*The Mechanical Euclid* (Cambridge 1837).

*On the Motion of Points Constrained and Resisted, and on the Motion of a Rigid Body. The Second Part of a new edition of A Treatise on Dynamics* (Cambridge 1834).

"On the Nature of the Truth of the Laws of Motion," *Transactions of the Cambridge Philosophical Society*, 5 (1834), 149–72.

*Novum Organon Renovatum* (London 1858), Pt. II of the 3rd ed. of *The Philosophy of the Inductive Sciences.*

*On the Philosophy of Discovery* (London 1860), Pt. III of the 3rd ed. of *The Philosophy of the Inductive Sciences.*

*The Philosophy of the Inductive Sciences*, 2nd ed. (London 1847), 2 vols.

*Of the Plurality of Worlds* (London 1853).

Review of Herschel's *Preliminary Discourse on the Study of Natural Philosophy, Quarterly Review*, 45 (1831), 374–407.

"Second Memoir on the Fundamental Antithesis of Philosophy," *Transactions of the Cambridge Philosophical Society*, 8 (1848), 614–20.

*Thoughts on the Study of Mathematics as Part of a Liberal Education* (Cambridge 1835).

"On the Transformation of Hypotheses in the History of Science," *Transactions of the Cambridge Philosophical Society,* 9 (1851), 139–47.

*The Collected Works of William Whewell,* G. Buchdahl and L. L. Laudan, eds, 10 vols. (London 1967–)

II. Selected Works Dealing with Whewell's Philosophy of Science

Blanché, Robert (1935): *Le Rationalism de Whewell,* Paris: Librairie Félix Alcan.

Buchdahl, Gerd (1971): "Inductivist versus Deductivist Approaches in the Philosophy of Science as Illustrated by some Controversies between Whewell and Mill," *The Monist,* 55, 3: 343–67.

Butts, Robert E. (1970): "Whewell on Newton's Rules of Philosophizing," in R. E. Butts and J. W. Davis, eds., *The Methodological Heritage of Newton,* Toronto: University of Toronto Press, pp. 132–49.

———. (1973): "Whewell's Logic of Induction," in R. N. Giere and R. S. Westfall, eds., *Foundations of Scientific Method: The Nineteenth Century,* Bloomington: Indiana University Press, pp. 53–85.

———. (1976): "William Whewell," in C. C. Gillispie, ed., *Dictionary of Scientific Biography,* Vol. XIV, New York: Scribners, pp. 292–95.

———. (1977): "Consilience of Inductions and the Problem of Conceptual Change in Science," in R. G. Colodny, ed., *Logic, Laws, and Life. Some Philosophical Complications* (University of Pittsburgh Series in the Philosophy of Science, Vol. 6), Pittsburgh: University of Pittsburgh Press, pp. 71–88.

———. (1987): "Pragmatism in Theories of Induction in the Victorian Era: Herschel, Whewell, Mach and Mill," in H. Stachowiak, ed., *Pragmatik. Handbuch Pragmatischen Denkens* (Bd. II, *Der Aufstieg pragmatischen Denkens im 19. und 20. Jahrhundert),* Hamburg: Felix Meiner Verlag, pp. 40–58.

De Morgan, Augustus (1859); Review of Whewell's *Novum Organon Renovatum, The Athenaeum,* No. 1682 (Jan. 8).

———. (1860): Review of Whewell's *Philosophy of Discovery, The Athenaeum,* No. 1694 (Apr. 14).

Ducasse, Curt J. (1960): "Whewell's Philosophy of Scientific Discovery," in E. H. Madden, ed., *Theories of Scientific Method: The Renaissance through the Nineteenth Century,* Seattle: University of Washington Press, pp. 183–217.

Fisch, Menachem (1985): "Whewell's Consilience of Inductions—An Evaluation," *Philosophy of Science,* 52, 2: 239–55.

Harper, William (1989): "Consilience and Natural Kind Reasoning," in J. R. Brown and J. Mittelstrass, ed., *An Intimate Relation. Studies in the History and Philosophy of Science Presented to Robert E. Butts on his 60th Birthday*, Dordrecht: Kluwer Academic Publishers.

Herschel, Sir J. F. W. (1841): Review of Whewell's *History and Philosophy of the Inductive Sciences, Quarterly Review*, No. 135 (June).

Hesse, Mary (1971): "Whewell's Consilience of Inductions and Predictions," *The Monist*, 55, 3: 520–24.

Laudan, Larry (1971): "William Whewell on the Consilience of Inductions," and "Reply to Mary Hesse," *The Monist*, 55, 3: 368–91, 525. Reprinted with revisions in Larry Laudan (1981): *Science and Hypothesis*, Dordrecht: D. Reidel, pp. 161–80.

Marcucci, Silvestro (1963): *L' "Idealismo" Scientifico di William Whewell*, Pisa: Istituto di Filosophia.

Niiniluoto, Ilkka (1978): "Notes on Popper as Follower of Whewell and Peirce," *Ajatus* 37 (Yearbook of the Philosophical Society of Finland): 272–327.

Ruse, Michael (1976): "The Scientific Methodology of William Whewell," *Centaurus*, 20, 3: 227–57.

von Wright, Georg H. (1957): *The Logical Problem of Induction*, 2nd rev. ed. Oxford: Oxford University Press, chs. III, IV.

# Notes

## INTRODUCTION

1. Two biographical studies of Whewell appeared in the nineteenth century. Issac Todhunter, *William Whewell, D. D., Master of Trinity College, Cambridge. An Account of his Writings with Selections from his Literary and Scientific Correspondence* (London 1876, 2 vols.) deals mainly with his scientific and philosophical work. Mrs. Stair Douglas, *The Life of William Whewell* (London 1881) discusses his personal and academic life. Both works contain numerous letters. Whewell's papers are in the Library of Trinity College.

2. This section of the Introduction is a slightly altered version of Parts II and III of my essay, "Necessary Truth in Whewell's Theory of Science," *American Philosophical Quarterly*, 2, 3 (July, 1965.) I am grateful to the editor and to the University of Pittsburgh Press for permission to use this material. See also my "On Walsh's Reading of Whewell's View of Necessity," *Philosophy of Science*, 32, 2 (April 1965). Additional references to discussions of Whewell's theory of necessity are given in the Bibliography.

3. William Whewell, *The Philosophy of the Inductive Sciences*, 2d. ed. (London 1847), vol I, 28.

4. William Whewell, *Novum Organon Renovatum*, (London 1858), 187. See below, p. 211.

5. In *The History of Scientific Ideas*, (London 1858), vol. I, 87, Whewell states that his discussion of space and time contains "the leading arguments respecting Space and Time, in Kant's *Kritik*." See also, *On the Philosophy of Discovery* (London 1860), 335. The problem of Whewell's debt to Kant has been studied in George C. Seward, *Die Theoretische Philosophie William Whewells und der Kantische Einfluss* (Tübingen 1938). However, Seward was so bent on making Whewell into a full-fledged Kantian that he missed the central problems. Silvestro Marcucci, *L' "Idealismo" Scientifico di William Whewell* (Pisa 1963) contains intriguing suggestions for the solution of the problem. The outline of my own view is given in "Professor Marcucci on Whewell's Idealism," *Philosophy of Science* 34, 2 (June 1967).

6. William Whewell, "Second Memoir on the Fundamental Antithesis of Philosophy," *Transactions of the Cambridge Philosophical Society*, vol. 7, pt. V, (1848), 33–35.

7. *Philosophy of Discovery*, 357–58.

8. William Whewell, "Demonstration that All Matter is Heavy," *Transactions of the Cambridge Philosophical Society*, vol. 7, pt II (1851).

9. See below, 139, 153 ff.

10. *History of Scientific Ideas*, vol. I, 58.

11. *Ibid.*, 217–18.

12. William Whewell, *On the Motion of Points Constrained and Resisted, and on the Motion of a Rigid Body. The Second Part of a new edition of A Treatise on Dynamics* (Cambridge 1834), vi.

13. William Whewell, *An Elementary Treatise on Mechanics*, 1st ed. (Cambridge 1819) and 5th ed. (Cambridge 1836), vii. Cf. *On the Motion of Points Constrained* . . . , v-vii. It is this second sense of "experience" that Whewell employs in answering some of his critics. For example, he writes of Mill: "Mr. Mill cannot deny that our knowledge of geometrical axioms and the like, *seems* to be *necessary*. I cannot deny that our knowledge, axiomatic as well as other, *never is* acquired *without experience*." *Philosophy of Discovery*, 286. See below, 304.

14. *Novum Organon Renovatum*, 59–63. See below, 130 ff.

15. See *Philosophy of the Inductive Sciences*, 276, where, when speaking of necessary truths, Whewell writes, "Truths *thus* necessarily acquired in the course of all experience, cannot be said to be learnt *from experience*, in the same sense in which particular facts at definite times, are learnt from experience, learnt by some persons and not by others, learnt with more or less of certainty."

16. *Ibid.*, 62.

17. *Philosophy of Discovery*, 342.

18. *History of Scientific Ideas*, vol. I, 59.

19. *Philosophy of Discovery*, 336–37.

20. *Ibid.*, 339.

21. *Ibid.*, 344.

22. *Ibid.*, 349–50.

23. Whewell, "On the Fundamental Antithesis of Philosophy," *Transactions of the Cambridge Philosophical Society*, vol. 7, Pt. II (1844), 27. See below, Ch. I.

24. What follows is extensively revised to incorporate my own better understanding of some features of Whewell's logic of induction. See Bibliography: Butts 1973, 1977, 1987; Niiniluoto 1978; Ruse 1976.

25. Whewell's Inductive Tables appear in Chapter III.

26. *Novum Organon Renovatum*, 70–71. See below 139.

27. Recent work by philosophers of science has pointed to a fundamental distinction between simplicity as the assertive richness of a conceptual scheme, and the simplicity of scientific laws. The former has to do with working out measures that will determine which among several schemes can, roughly speaking, express more. The less complex scheme in this respect will be simpler. The simplicity of laws, on the other hand, depends on the conceptual richness of predicates, and also on the form and content of the laws. I think Whewell was working out the rudiments of determining this second kind of simplicity. See *Novum Organon Renovatum*, 85–96, 195–232 (also below, 151–160, 218–248). Certainly he was on to something important here. He would have agreed completely with Goodman's recent statement, "No longer do we need to take seriously the idea that simplicity is something to be sought only if there is time after truth has been attained. To seek truth is to seek a true system, and to seek system at all is to seek simplicity. And considerations of simplicity are inextricably involved in the selection of what is regarded as a true system." Nelson Goodman, "Axiomatic Measurement of Simplicity," *The Journal of Philosophy*, LII, 24 (Nov. 24, 1955), 709.

28. *Novum Organon Renovatum*, 108 (below 170).

29. *Ibid.*, 110 (below 171–172). Whewell unfortunately uses the term "induction" ambiguously. Most appropriately, for him, it refers to "the *process* of a true Colligation of Facts by means of an exact and appropriate Conception." (*Ibid.*, 70; below 138) And this process, though sometimes he calls it a kind of inferring or reasoning, "is a leap which is out of the reach of method." (*Ibid.*, 114; below 175) He also uses the term to denote the *proposition* resulting from the colligating process. (*Ibid.*, 70; below 138). The ambiguity, as will become apparent shortly, makes it difficult to assess Whewell's view on induction as a special form of inference.

30. *Ibid.*, 70 ff, Bk. III (below 138 ff, 191 ff).

31. *Ibid.*, 105 (below 167).

32. *Ibid.*, 106 (below 168–169).

33. *Ibid.*, 107 (below 169).

34. *Ibid.*, 115 (below 175–176).

35. *History of Scientific Ideas*, I, 248–49.

36. *Novum Organon Renovatum*, 112 (below 173).

37. *Ibid.*, 113 (below 174).

38. *Ibid.*, 112 (below 173).

39. *History of Scientific Ideas*, I, 260–61.

40. There is much more to the story than I am here telling. As I think I have shown conclusively in "Necessary Truth in Whewell's Theory of Science," *American Philosophical Quarterly*, 2, 3 (July 1965), 13–19. Whewell did make some attempt on theological and metaphysical grounds to justify the necessary truths at which he thinks all sciences aim.

41. *Novum Organon Renovatum*, 115 (below 176).

42. See, for example, H. L. Mansel, *Prolegomena Logica. An Inquiry into the Psychological Character of Logical Processes* (Oxford 1851). Even George Boole entitled his major work in logic *An Investigation into the Laws of Thought* (London, 1854).

43. More can be said about the pragmatic aspects of Whewell's theory of science, his views on the kinds of *decisions* that are relevant in methodology. See Bibliography, Butts 1987.

44. *Novum Organon Renovatum* (below 151–53).

45. *Ibid.*, (below 153).

46. At present the most discussed feature of Whewell's theory of induction is the idea of consilience. See Bibliography: Butts 1970, 1973, 1977, 1987; Fisch 1985; Harper 1989; Hesse 1971; Laudan 1971, 1981; Niiniluoto 1978. Peter Achinstein and Malcolm Forster (and perhaps others) have work in progress on related issues. The earlier papers by Hesse, Laudan, and Butts tried to provide formal models for consilience. There is, of course, much disagreement amongst these authors concerning which theory of confirmation is most faithful to Whewell's intentions. Harper's paper is an elegant study of theoretical unification in Newton's physical synthesis; for that reason it should be read together with Whewell's own use of Newton's physics as the chief illustration of his own theory of method.

47. "Of the Transformation of Hypotheses in the History of Science" (below 262).

## CHAPTER I

### REMARKS ON MATHEMATICAL REASONING AND ON THE LOGIC OF INDUCTION

1. Dugald Stewart, *Elements of the Philosophy of the Human Mind,* II (London and Edinburgh 1814).

2. *Ibid.,* 38.

3. William Whewell, *Thoughts on the Study of Mathematics as a Part of a Liberal Education,* (Cambridge 1835).

4. Pappus, *Pappi Alexandrinni Mathematicae Collectiones,* (Venice 1589), B.VIII, Prop. IX. I purposely omit the confusion produced by this author's mode of treating the question, in which he inquires the force which will *draw* a body *up* the inclined plane.

5. See William Whewell, *History of the Inductive Sciences,* II, (London 1847), 14–18, "Revival of the Scientific Idea of Pressure. Stevinus. Equilibrium of Oblique Forces."

### ON THE FUNDAMENTAL ANTITHESIS OF PHILOSOPHY

1. William Whewell, *History of Scientific Ideas,* (London 1858), 165–85.

## CHAPTER II

1. The accelerative quantity of a force (the *quantitas acceleratrix vis cujusvis* of Newton) is often called *the accelerating force;* and we may thus have to speak of *the accelerating force of a certain force,* which is at any rate an awkward phraseology. It would perhaps have been fortunate if Newton, or some other writer of authority, at the time when the principles of mechanics were first clearly developed, had invented an abstract term for this quantity: it might for instance have been called *accelerativity.* And the second law of motion would then have been, that the *accelerativity* of the same force is the same, whatever be the motion of the body acted on.

2. The motive quantity of a force (*vis cujusvis quantitas motrix* of Newton) is sometimes called *moving force;* we are thus led to speak of the moving force of a force, as we have already observed concerning accelerating force. Hence, as in that case, we might employ a single term, as *motivity,* to denote this property of force; and might thus speak of it and of its measures without the awkwardness which arises from the usual phrase.

## CHAPTER III

### NOVUM ORGANON RENOVATUM

1. William Whewell, *The History of Scientific Ideas,* I, Bk. II, Ch. VI, sect. 6.

2. *Ibid.,* Bks. V, VI, VII, VIII, IX, X.

3. Aristotle, *Analytica Priora,* 68b15–20.

4. The syllogism here alluded to would be this:—

Mercury, Venus, Mars, describe ellipses about the Sun;
All Planets do what Mercury, Venus, Mars, do;
Therefore all Planets describe ellipses about the Sun.

5. Εἰ οὖν ἀντιστρέφει τὸ Γ τῷ Β καὶ μὴ υπερτείνει τὸ μέσον.—Aristotle. *Ibid.*

6. I here take the liberty of characterizing inventive minds in general in the same phraseology which, in the *History of the Inductive Sciences,* I have employed in references to particular examples. These expressions are what I have used in speaking of the discoveries of Copernicus. *History of the Inductive Sciences,* I, 388–400.

7. These observations are made on occasion of Kepler's speculation, and are illustrated by references to his discoveries. *Ibid.*, I, 432 ff.

8. *Ibid.*, I, 434.

9. *Ibid.*, II, 168.

10. Jean André Deluc, *Recherches sur les Modifications de l'Atmosphère,* (Paris 1784), Partie I.

11. Whewell, *History of Inductive Sciences,* II, 451–61.

12. *Ibid.*, II, 472.

13. *Ibid.*, II, 458–61.

14. *Ibid.*, II, 489.

15. *Ibid.*, II, 463.

16. *Ibid.*, II, 442.

17. *Ibid.*, II, 461–66.

18. *Ibid.*, II, 583–88.

19. In the Inductive Tables they are marked by an asterisk.

20. Whewell, *History of the Inductive Sciences,* III, 37–40.

21. Auguste Comte, *Cours de Philosophie Positive,* (Paris 1830–1842).

22. Whewell, *History of the Inductive Sciences,* III, 468–74.

23. *Ibid.*, III, 399.

24. *Ibid.*, I, 210–18.

25. *Ibid.*, II, 273.

26. On the precautions employed in astronomical instruments for the measure of space, see Sir John Herschel, *Treatise on Astronomy,* (London 1833), articles 103–110.

27. On the precautions employed in the measure of time by astronomers, see *ibid.*, articles 115–127.

28. A *meridian* is a circle passing through the poles about which the celestial sphere revolves. The meridian *of any place* on the earth is that meridian which is exactly over the place.

29. William Whewell, *The Philosophy of the Inductive Sciences,* 2nd ed. (London 1847), I, 319–44.

30. Michael Faraday, *Chemical Manipulation,* 3rd. ed. (London 1842), 3.

31. Whewell, *History of the Inductive Sciences,* III, 340–41.

32. [See this volume 133.]

33. *History of the Inductive Sciences,* I, 200.

34. The formula is $v = 2{,}037\ (a^t - 1)$ where $v$ is the velocity of cooling, $t$ the temperature of the thermometer expressed in degrees, and $a$ is the quantity 1,0077.

The degree of coincidence is as follows:—

| Excess of temperature of the thermometer, or values of $t$. | Observed values of $v$. | Calculated values of $v$. |
|---|---|---|
| 240 | 10·69 | 10·68 |
| 220 | 8·81 | 8·89 |
| 200 | 7·40 | 7·34 |
| 180 | 6·10 | 6·03 |
| 160 | 4·89 | 4·87 |
| 140 | 3·88 | 3·89 |
| 120 | 3·02 | 3·05 |
| 100 | 2·30 | 2·33 |
| 80 | 1·74 | 1·72 |

35. For if $\theta$ be the temperature of the inclosure, and $t$ the excess of temperature of the hot body, it appears, by this law, that the radiation of heat is as $a\theta$. And hence the quickness of cooling, which is as the excess of radiation, is as $a^{\theta+t}—a^{\theta}$; that is, as $a\theta$ $(a^{t}—1)$ which agrees with the formula given in the last note.

The whole of this series of researches of Dulong and Petit is full of the most beautiful and instructive artifices for the construction of the proper formulæ in physical research.

36. [See *Nikolaus Kapernikus Gesamtausgabe,* II, (München, 1949), 12.]

37. Whewell, *History of the Inductive Sciences,* II, 411–12.

38. Sir John Herschel, "On the Investigation of the Orbits of Revolving Double Stars," *Memoirs of the Royal Astronomical Society,* V (1833), 171–222, 179.

39. *Ibid.,* 180.

40. *Ibid.,* 181.

41. Provided the argument of the law which we neglect have no coincidence with the argument of the law which we would determine.

42. J. W. Lubbock, "On the Tides in the Port of London," *Philosophical Transactions of the Royal Society of London,* pt. II (1831), 379–415.

43. This period of nineteen years was also selected for a reason which is alluded to in a former note. It was thought that this period secured the inquirer from the errours which might be produced by the partial coincidence of the Arguments of different irregularities; for example, those due to the moon's Parallax and to the moon's Declination. It has since been found (William Whewell, "On the Determination of the Laws of the Tides from Short Series of Observations," *Philosophical Transactions of the Royal Society of London* [1838]), that with regard to Parallax at least, the Means of one year give sufficient accuracy.

44. Galileo Galilei, *Dialogue Concerning the Two Chief World Systems,* trans. Stillman Drake (Berkeley and Los Angeles 1953), 20.

45. Gottfried Leibniz, *Opera Omnia,* Louis Dutens, ed., I (Geneva 1768), 366.

46. E. S. Haldane and G. R. T. Ross, *The Philosophical Works of Descartes,* I (Cambridge 1931), 267.

47. [See John Playfair, "Dissertation on the Progress of Mathematical and

Physical Sciences since the Revival of Learning in Europe," in *Encyclopaedia Britannica,* (Supplement to 4th, 5th, and 6th editions).]

48. Florian Cajori, (tr.), *Sir Isaac Newton's Mathematical Principles of Natural Philosophy and his System of the World* (Berkeley 1946), 411.

49. *Ibid.,* 413.

50. Michael Faraday, *Experimental Researches in Electricity,* I (New York 1965), 420.

51. These words refer to another proposition, also established by the Method of Gradation.

52. Whewell, *History of the Inductive Sciences,* III, 181–92.

53. Faraday, *Experimental Researches in Electricity,* 142.

54. J. M. Robertson, *The Philosophical Works of Francis Bacon* (London 1905), 328.

55. Whewell, *History of the Inductive Sciences,* III, 554–56.

56. *Ibid.,* III, 580–87.

OF THE TRANSFORMATION OF HYPOTHESES IN THE HISTORY OF SCIENCE

1. E. S. Haldane and G. R. T. Ross, *The Philosophical Works of Descartes,* I, 281.

2. M. Jean Bernoulli, *Nouvelles Pensées sur le Système de M. Descartes* (Paris 1689), 239.

3. Christian Huyghens, *Traité de la Lumière. . . . Avec un Discourse de la Cause de la Pesanteur* (Leyde 1690), 135.

4. M. Saurin, "D'une difficulté considerable proposée par M. Hughens contre le Systême Cartesien sur la cause de la Pesanteur," *Histoire de L'Académie Royale Des Sciences, Avec les Mémoires* (Paris 1709), 131–49. Bulfinger . . . conceived that by making a sphere revolve at the same time about two axes at right angles to each other, every particle would describe a great circle; but this is not so.

5. M. Saulmon, "Quadrature d'une Zone circulaire," *Histoire de L'Académie Royale Des Sciences, Avec les Mémories* (Paris 1714), 156–74.

6. Abbé Joseph Privat de Molières, "Les Loix Astronomiques des vîtesses des Planetes dans leur Orbes, expliquées méchaniquement dans le System du Plein," *Histoire L'Académie Royale Des Sciences, Avec les Mémoires* (Paris 1733), 301–12.

7. . . . If we abandon the clear principles of mechanics, the writer says, "toute la lumière que nous pouvons avoir est éteinte, et nous voilà replongés de nouveau dans les anciennes tenébres du Peripatetisme, dont le Ciel nous veuille preserver!" It was also objected to the Newtonian system, that it did not account for the remarkable facts, that all the motions of the primary planets, all the motions of the satellites, and all the motions of rotation, including that of the sun, are in the same plane; facts which have been urged by Laplace as so strongly recommending the Nebular Hypothesis; and that hypothesis is, in truth, a hypothesis of vortices respecting the *origin* of the system of the world.

8. Bernoulli, John, *Nouvelle Physique Celeste* (Paris n. d.), 163. The deviation of the orbits of the planets from the plane of the sun's equator was of course a difficulty in the system which supposed that they were carried round by

the vortices which the sun's rotation caused, or at least rendered evident. Bernoulli's explanation consists in supposing the planets to have a sort of *leeway (dérive des vaisseaux)* in the stream of the vortex.

9. Whewell, *The History of Scientific Ideas,* Bk. III, Ch. IX, Art. 7.

10. See J. S. Mill, *A System of Logic,* 2nd ed., I, 311.

## CHAPTER IV

1. [All Whewell's references are to the first edition of 1843, *A System of Logic, Ratiocinative and Inductive, being a connected view of the Principles of Evidence, and of the Methods of Scientific Investigation* by John Stuart Mill, unless otherwise indicated.]

2. On this subject see an Essay, "On the Transformation of Hypotheses . . . ," above, ch. III.

3. Whewell, *History of the Inductive Sciences,* II, 206–07.

4. Whewell, *Philosophy of the Inductive Sciences,* I, 259–60.

5. *Ibid.,* I, 51–53.

6. J. S. Mill, *A System of Logic,* Bk. III, ch. VIII.

7. Sir John Herschel, *A Preliminary Discourse on the Study of Natural Philosophy* (London 1831), 183–84.

8. Whewell, *History of the Inductive Sciences,* III, 21–26.

9. Whewell, *Philosophy of the Inductive Sciences,* II, 429.

10. Whewell, *History of the Inductive Sciences,* III, 105–07.

11. Given also in Whewell, *Philosophy of the Inductive Sciences,* II, 410–11.

12. *Ibid.,* I, 412 ff.

13. See Whewell, *History of the Inductive Sciences,* III, 67–76.

14. There are some points in my doctrines on the subject of the Classificatory Sciences to which Mr. Mill objects, (ii. 314, &c.), but there is nothing which I think it necessary to remark here, except one point. After speaking of Classification of organized beings in general, Mr. Mill notices (ii. 321) as an additional subject, the arrangement of natural groups into a Natural Series; and he says, that "all who have attempted a theory of natural arrangement, including among the rest Mr. Whewell, have stopped short of this: all except M. Comte." On this I have to observe, that I stopped short of, or rather passed by, the doctrine of a Series of organized beings, because I thought it bad and narrow philosophy: and that I sufficiently indicated that I did this. In the *History of the Inductive Sciences,* III, 389, I have spoken of the doctrine of Circular Progression propounded by Mr. Macleay, and have said, "so far as this view *negatives* a mere *linear* progression in nature, which would place each genus in contact with the preceding and succeeding ones, and so far as it requires us to attend to the more varied and ramified resemblances, there can be no doubt that it is supported by the result of all the attempts to form natural systems." And with regard to the difference between Cuvier and M. de Blainville, to which Mr. Mill refers (ii. 321), I certainly cannot think that M. Comte's suffrage can add any weight to the opinion of either of those great naturalists.

15. Whewell, *History of the Inductive Sciences,* II, 601–02.

16. Whewell, *Philosophy of the Inductive Sciences,* II, 65–68.

17. I have given elsewhere [*Philosophy of Discovery*, ch. XXI] reasons why I cannot assign to M. Comte's *Philosophie Positive* any great value as a contribution to the philosophy of science. In this judgment I conceive that I am supported by the best philosophers of our time. M. Comte owes, I think, much of the notice which has been given to him to his including, as Mr. Mill does, the science of society and of human nature in his scheme, and to his boldness in dealing with these. He appears to have been received with deference as a mathematician: but Sir John Herschel has shown that a supposed astronomical discovery of his is a mere assumption. I conceive that I have shown that his representation of the history of science is erroneous, both in its details and in its generalities. His distinction of the three stages of sciences, the theological, metaphysical, and positive, is not at all supported by the facts of scientific history. Real discoveries always involve what he calls *metaphysics;* and the doctrine of final causes in physiology, the main element of science which can properly be called *theological,* is retained at the end, as well as the beginning of the science, by all except a peculiar school.

18. I have also, in the same place, given the Inductive Pyramid for the science of Optics. These Pyramids are necessarily inverted in their form, in order that, in reading in the ordinary way, we may proceed *to* the vertex. Whewell, *Philosophy of the Inductive Sciences,* II, insert between 118 and 119. [See Tables in Chapter III, this volume].

19. Fredrich Alexander von Heinrich Humboldt, *Cosmos, Sketch of a Physical Description of the Universe,* Translated under the direction of E. C. Otté, II (London 1849–58), note 35.

20. The reader will probably recollect that as *Induction* means the inference of general propositions from particular cases, *Deduction* means the inference by the application of general propositions to particular cases, and by combining such applications; as when from the most general principles of Geometry or of Mechanics, we prove some less general theorem; for instance, the number of the possible regular solids, or the principle of *vis viva.*

21. Whewell, *History of the Inductive Sciences,* II, 80–91.

22. *Ibid.,* II, 92–132.

23. *Ibid.,* II, 119.

24. *Ibid.,* II, 357–68.

25. [See this volume chap. I.]

26. Sir John Herschel, Review of the *History and Philosophy of the Inductive Sciences, Quarterly Review,* No. 135 (June 1841), 96–130.

## CHAPTER V

### NEWTON

1. Sir Isaac Newton, *Opticks* (New York 1952), 404.
2. *Ibid.,* 369.
3. Whewell, *History of the Inductive Sciences,* I, 443–52, and II, 160–94.
4. Newton, *Opticks,* 404.
5. Whewell, *The History of Scientific Ideas,* Bk. III, ch. X.

6. *Ibid.*, Bk. III, chs. IX, X, XI.

7. Newton, *Optics*, 401–02.

8. J. Spedding, R. L. Ellis, D. D. Heath, (eds.), *The Works of Francis Bacon*, I (London 1858), 229. "Licet enim in natura nihil existet præter corpora individua, edentia actus puros individuos ex lege; in doctrinis tamen illa ipsa lex, ejusque inquisitio, et inventio, et explicatio, pro fundamento est tam ad sciendum quam ad operandum. Eam autem *legem*, ejusque *paragraphos*, *formarum* nomine intelligimus; præsertim cum hoc vocabulum invaluerit, et familiariter occurrat."

*Ibid.*, 258. "Eadem res est *forma* calidi vel *forma* luminis, et *lex* calidi aut *lex* luminis."

# Index